DARWIN, GOD AND THE MEANING OF LIFE

If you accept evolutionary theory, can you also believe in God? Are human beings superior to other animals, or is this just a human prejudice? Does Darwin have implications for heated issues like euthanasia and animal rights? Does evolution tell us the purpose of life, or does it imply that life has no ultimate purpose? Does evolution tell us what is morally right and wrong, or does it imply that ultimately *nothing* is right or wrong? In this fascinating and intriguing book, Steve Stewart-Williams addresses these and other fundamental philosophical questions raised by evolutionary theory and the exciting new field of evolutionary psychology. Drawing on biology, psychology and philosophy, he argues that Darwinian science supports a view of a godless universe devoid of ultimate purpose or moral structure, but that we can still live a good life and a happy life within the confines of this view.

STEVE STEWART-WILLIAMS is a lecturer in evolutionary psychology at Swansea University. Before taking this position, he completed his PhD at Massey University in New Zealand, and then did a postdoctoral fellowship at McMaster University in Canada. His research and writing cover a diverse range of topics, including the placebo effect, the philosophy of biology, and the evolution of altruism and mating behaviour. His hobbies include playing the guitar and solving the riddle of the meaning of life.

Darwin, God and the Meaning of Life

HOW EVOLUTIONARY THEORY UNDERMINES EVERYTHING YOU THOUGHT YOU KNEW

Steve Stewart-Williams

CAMBRIDGE
UNIVERSITY PRESS

CAMBRIDGE UNIVERSITY PRESS

Cambridge, New York, Melbourne, Madrid, Cape Town, Singapore, São Paulo,
Delhi, Dubai, Tokyo, Mexico City

Cambridge University Press
The Edinburgh Building, Cambridge CB2 8RU, UK

Published in the United States of America by Cambridge University Press, New York

www.cambridge.org
Information on this title: www.cambridge.org/9780521762786

First published 2010

Printed in the United Kingdom at the University Press, Cambridge

A catalogue record for this publication is available from the British Library

Library of Congress Cataloguing in Publication data
Stewart-Williams, Steve, 1971–
Darwin, God and the meaning of life : how evolutionary theory undermines
everything you thought you knew / Steve Stewart-Williams.
p. cm.
Includes bibliographical references and index.
ISBN 978-0-521-76278-6   1. Evolution (Biology) – Philosophy.   2. Evolution (Biology) –
Religious aspects.   3. Evolution (Biology) – Moral and ethical aspects.   I. Title.
QH360.5.S79 2010
231.7'652–dc22
2010021885

ISBN 978-0-521-76278-6 Hardback

For Jane

Contents

Acknowledgments

My main reason for writing this book is that it's the book I've wanted to *read* for the last decade or so. But I couldn't have done it alone. Right from the start, I was aided and abetted by some of the best friends and best colleagues I could ever have hoped for. At the top of the list is my wife, the mother of my children, the love of my life: Jane Stewart-Williams. I am also indebted to friends and colleagues in each of the three countries in which this book was written: New Zealand, Canada, and Wales. In New Zealand, I especially want to thank my PhD supervisors, John Podd, Stephen Hill, and James Battye. Thanks in particular to Poddy, who encouraged my decision to change my PhD topic a year into the doctoral programme. It was a risky move but it worked out well in the end, and if I hadn't done it, I might never have been in the position to write this book (long story). Thanks to my post-doc mentors in Canada, Martin Daly and Margo Wilson, for agreeing to host my post-doc on only a few weeks' notice (also a long story); for taking it in their stride when I accidentally set their kitchen on fire; and for being brilliant, brilliant scientists. Sadly, Margo died recently, which is a great personal loss to me and many others, and also a great loss to science. So thanks again to her and to Martin, and to all the folks in the Daly–Wilson lab. Thanks also to my colleagues at Swansea University for providing an extremely supportive environment as I turned an oversized collection of notes and musings into the book you're holding in your hands.

I owe a big debt of gratitude to everyone who proofread chapters of the book or the entire manuscript: Giles Boutel, Stephen Hill, Danny Krupp, Rob Lowe, Jane Stewart-Williams, Andrew Thomas, J.J. Tomash, and my mum, Bea Stewart. A special thanks to Giles Boutel; Giles and I have been encouraging each other's writing efforts since we were nine years old. Thanks to Andrew Peart at Cambridge University Press for knocking on my office door and offering me a book contract, and to Hetty Reid who took over the reins when Andrew left for another job. Thanks to Claire Williams for helping out with the book cover. And thanks to Mark Ayling, Hilary Nicholl Bartle, Alan Beaton, Laura Bell, Dave Benton, Mark Blagrove, Stephen Boulter, Paul Brain, Brian Cainen, Gretchen Carroll, Neil Carter, Julia Davis-Coombs, Gerald Dreaver, Daniel Dunlavey, Simon Dymond, Jason Evans, Laura Finn, Nikki Finnemore, Iain Ford, Anna Harrod, Marcus Hinds, Daniel Jeffares, Michelle Lee, Andrew Loughnan, Daniel Loughnan, Lisa Osborne, Phil Newton, Carrie Parkinson, Andy Parrott, Ian Pepper, Richard Porter, Sandanalakshmi Ramakrishnan, Jordan Randell, Phil Reed, Ian Russell, Tim Ryder, Jackie Scholz, David Skibinski, Dan Skipper, Paul Spence, Jolyon Stewart, Lynsey Thomas, Phil Tucker, Sophie Ulbrich-Ford, Barbara Williams, and Brian Williams for miscellaneous help and input. Last but not least, thanks to Darwin and India for putting up with a sometimes-grumpy daddy as he toiled away on his little project.

ONE

Darwin and the big questions

It may well be that for posterity his name will stand as a turning point in the intellectual development of our western civilization ... If he was right, men will have to date from 1859 the beginning of modern thought.

Will Durant (1931), p. 22

Because all organisms have descended from a common ancestor, it is correct to say that the biosphere as a whole began to think when humanity was born. If the rest of life is the body, we are the mind.

E. O. Wilson (2002), p. 132

I simply can't stand a view limited to this earth, I feel life is so small unless it has windows into other worlds. I feel it vehemently and instinctively and with my whole being.

Bertrand Russell, cited in Ray Monk (1996), p. 248

Why are we here?

Evolutionary theory answers one of the most profound and fundamental questions human beings have ever asked themselves, a question that has plagued reflective minds for as long as reflective minds have existed in the universe: why are we here? How did we come to exist on this planet? In a lot of ways, this is a very ordinary planet. It orbits an ordinary star and is part of an otherwise ordinary solar system, which is part of an ordinary galaxy, one of billions of galaxies in the visible universe. But in another way, this planet is very strange, because a small proportion of its surface has somehow transformed itself into *life*. Even stranger, a small proportion of the life forms on the planet are conscious of their own existence, and able to comprehend – at least to some degree – the world around them. How did this happen? How did tiny fragments of the universe come to be organized in such a way that they became conscious of themselves and their little corner of the world? Why, in other words, are we here?

The answer, first revealed by the English scientist Charles Darwin, is that we are here because we evolved. Not everyone accepts this answer, though. Darwin's theory of evolution by natural selection is one of those ideas that can divide a room. On the one hand, a lot people simply love it. It excites them, captivates them, and illuminates their understanding of the world. On the other hand, there are lots of people who really, *really* hate the idea. They think it's poisonous and socially corrosive. They view the Darwinian worldview as cold and deeply disturbing; one commentator described it as 'a dogma of darkness and death'.[1] Some even believe that evolutionary theory is part of a great conspiracy, designed to steer people away from God and push them instead towards atheism. But one thing that both friends and enemies of the theory would have to agree on is that, for better or for worse, Darwin's theory is one of the most important ideas in intellectual history. In fact, a lot of people suggest that it is *the* most important idea. And one reason it's so important is that it doesn't just have implications for esoteric, academic questions – the kind of questions that are only of interest to a tiny minority of intellectuals. Unlike, say, atomic theory or relativity theory, the theory of evolution has important implications for how we view ourselves and our place in the universe. It has implications for things that are truly important to people, things that people care deeply about, things that most of us have an opinion about. Those are the things we'll explore in this book.

[1] Bryan (1925).

More specifically, we'll look at the implications of evolutionary theory for the following 'hot topics' in philosophy:

1. evolution and the existence of God;
2. evolution and the place of humankind in nature; and
3. evolution and morality (the question of what's right and what's wrong).[2]

At first glance, it might not seem that Darwin's theory – or any scientific theory, for that matter – would have implications for any topic in philosophy. Science deals with the empirical world and with matters of fact, whereas philosophy, almost by definition, deals with matters outside the remit of science. Within its domain of applicability, evolutionary theory is a striking success. But the theory does not have any obvious or immediate implications for any of the questions dealt with by philosophers. Indeed, some suggest that it has no philosophical implications at all. Some argue, for instance, that unless you're a Creationist or biblical literalist, evolutionary theory has no implications for the question of God's existence; it is compatible with atheism, certainly, but it is also compatible with theism. Others argue that evolutionary theory has no implications for questions of morality, because evolution deals with facts, whereas ethics deals with values, and you can't derive values from facts. The enigmatic Ludwig Wittgenstein summed up this general position when he noted that 'Darwin's theory has no more to do with philosophy than any other hypothesis in natural science'.[3]

Many disagree with this assessment, however, and for good reason. If nothing else, our traditional philosophical convictions look very different through the lens of evolutionary biology. As the great English philosopher Bertrand Russell wrote:

[2] Evolutionary theory also has important implications for epistemology – the philosophical analysis of knowledge – but I won't deal with that topic here. See Boulter (2007); Bradie (1986); Campbell (1974); Carruthers (1992); Hahlweg and Hooker (1989); Hull (2001); Lorenz (1941); O'Hear (1997); Plotkin (1993); Radnitzky and Bartley III (1987); Ruse (1986); Stewart-Williams (2005).
[3] Wittgenstein (1921), 4.1122.

If man was evolved by insensible gradations from lower forms of life, a number of things became very difficult to understand. At what moment in evolution did our ancestors acquire free will? At what stage in the long journey from the amoeba did they begin to have immortal souls? When did they first become capable of the kinds of wickedness that would justify a benevolent Creator in sending them into eternal torment? Most people felt that such punishment would be hard on monkeys, in spite of their propensity for throwing coconuts at the heads of Europeans. But how about *Pithecanthropus erectus*? Was it really he who ate the apple? Or was it *Homo pekinensis*?[4]

These are awkward questions, and there are plenty more where they came from. Why would the omnipotent creator of the entire universe be so deeply attached to a bipedal, tropical ape? Why would he take on the bodily form of one of these peculiar tailless primates? Why would such a magnificent being be so obsessively, nit-pickingly preoccupied with trivial matters such as the dress code and sexual behaviour of one mammalian species, especially its female members? What are angels? Did they evolve through natural selection? Are they primate cousins of *Homo sapiens* like Neanderthals? If not, why do they look so much like us?

Questions such as these begin to signal the threat that evolutionary theory poses to our traditional philosophical convictions. However, the implications go far beyond raising awkward questions. Right from the start, people sensed that Darwin's theory would have radical implications for philosophy. Darwin himself suspected that this was the case. In one of his early notebooks on the transmutation of species, he made a cryptic remark: 'Origin of man now proved. Metaphysics must flourish. He who understands baboon would do more toward metaphysics than Locke.'[5] Being a rather cautious individual, he didn't elaborate much on this suggestion. Others, however, were not so reticent. In the latter half of the nineteenth century, there was a

---

[4] Russell (1950), p. 146. Note that *Pithecanthropus erectus* (or Java man) is now classed as a specimen of *Homo erectus*. *Homo pekinensis* (known also as Peking man or Beijing man) belongs to a subspecies of *Homo erectus*: *Homo erectus pekinensis*.

[5] Cited in Barrett *et al.* (1987), p. 539.

flood of suggestions about what Darwin's theory implied. There was also a lot of disagreement. (This shouldn't be too surprising; as the psychologist William James pointed out, the only thing philosophers can be relied on to do is to disagree with each other.) Some thought that evolutionary theory posed a grave threat to theistic belief, others that it was 'an advance in our theological thinking'.[6] Some thought the theory toppled us from our assumed perch at the top of the animal kingdom, others that it explained how and why we came to be superior among the animals. Some thought that evolution showed us what is right by moving towards it, others that the theory implied that *nothing* is right or wrong – in other words, that evolution undermined morality altogether. Within a decade of his death, Josiah Royce was able to write of Darwin's masterpiece, *On the Origin of Species by Means of Natural Selection*, that, 'With the one exception of Newton's *Principia*, no single book of empirical science has ever been of more importance to philosophy than this work of Darwin.'[7]

Unfortunately, though, for most of the twentieth century, philosophers all but forgot about Charles Darwin.[8] Most adopted the Wittgensteinian view that evolutionary theory had no implications for their field. Darwin's fortunes in philosophical circles didn't start changing until the latter decades of the twentieth century, when philosophers such as Daniel Dennett, Michael Ruse, and Peter Singer brought a deep appreciation of evolutionary theory to their work. Dennett in particular has made it his mission to sing Darwin's praises. In his view, Darwinism is a 'universal acid', the influence of which seeps out to infect every area of human thought. Philosophy, he argues, is no exception.

Of course, philosophy itself already enjoys a reputation as a powerful corrosive agent, challenging and overthrowing our most fundamental beliefs. The dyspeptic nineteenth-century philosopher Friedrich

---

[6] Cited in White (1896), p. 103.

[7] Royce (1892), p. 286.

[8] Cunningham (1996).

Nietzsche (who was strongly influenced by Darwin's ideas) described it as 'a terrible explosive from which nothing is safe'. Thus, the question we'll be concerned with in this book is: what happens when you mix a universal acid with a terrible explosive? The short answer is that it challenges some of our most cherished and longest-standing beliefs about God, man, and morality. The long answer is the remainder of this book. But before we launch into a detailed discussion of any of these issues, let's take a brief tour of the terrain we'll be covering, the questions we'll be asking, and the opinions we'll be meeting along the way.

Part I: Darwin gets religion[9]

I open with the question of the existence of God. Did God create us in his image, or did we create him in ours? Or, as Nietzsche put it, 'Is man one of God's blunders? Or is God one of man's blunders?' Here are some of the other questions we'll be asking in this section:

> Can someone who believes in evolution believe in God as well? • Did God directly guide the evolutionary process? • Did God choose natural selection as his means of creating life indirectly? • Must we invoke God to explain the origin of life, the origin of the universe, or the origin of consciousness? • Does the suffering entailed by natural selection suggest that there could be no God – or that if there is a God, he must be evil?

We won't be concerned with the full range of arguments for and against God's existence, but only those directly related to Darwin's theory. Here's a synopsis of the chapters that make up Part I (Chapters 2–7).

Chapter 2: Clash of the Titans

We'll start at the beginning: did we evolve or did God create us in our present form? These are not, of course, the only options, but they're the ones that get the most airtime and that are the most important to

---

[9] Parts of this section are based on Stewart-Williams (2004c).

the most people. One of them also happens to be correct. Those readers already familiar with evolutionary theory, and with the evolution v. creationism debate, may wish to skip ahead to the next chapter, but my hope is that even these readers will find something of value here. The chapter starts with a sketch of the Creationist viewpoint followed by a sketch of evolutionary theory. (Did you know that Darwin did not actually originate the concept of evolution?) Then we'll survey some of the fascinating and bizarre evidence supporting Darwin's theory. We'll see how evolutionary theory explains otherwise inexplicable facts about the biological world, such as why bat wings are less like bird wings than they are like whale flippers; why flightless birds have wings; why human embryos have gill slits; why whales are occasionally born with hind limbs; and why humans are occasionally born with tails. Finally, we'll examine some of the arguments against evolutionary theory. One of the most persuasive of these asserts that there are certain things in the biological world that simply could not have evolved through natural selection. This includes the bacterial flagellum, the immune system, and the blood-clotting system. Even some scientifically minded laypeople are secretly given pause by these apparently reasonable arguments. If you're one of them, you've been taken in by the slick marketing of the Intelligent Design movement. Hopefully, by the end of Chapter 2, you'll never suffer from this malady again.

Chapter 3: Design after Darwin

In Chapter 3, we'll turn our attention to an important philosophical argument for the existence of God: the argument from design. The basic idea is as follows. Certain parts of the natural world look as though they were designed (eyes, teeth and claws, for example); you can't have design without a designer; thus, there must be a designer and the designer is God. We'll see that evolutionary theory undermines this argument and therefore poses a serious threat to theistic belief, *even for those who believe for other reasons.* We'll also see that, within this area of philosophy, Darwin had a greater impact than one of the greatest philosophers of all time: David Hume.

Chapter 4: Darwin's God

The remaining chapters of Part I address the question: can someone who believes in evolution also believe in God? In a sense, the answer is clearly yes, you can believe in both. We know this because a lot of people *do* believe in both evolution and God. This fact is often held up as evidence that these beliefs must be compatible. But the conclusion is too strong; it shows only that, if they're incompatible, it's not in any obvious or straightforward way. We might still find that, when we look more closely, they are inconsistent with one another. In addressing this issue, we'll sample some of the clever ways that believers have tried to meld their belief in God with the truth of evolution. Some suggest, for instance, that God personally guided the evolutionary process, either in whole or in part. Others suggest that, rather than intervening, God chose natural selection as his means of creating life. Many notable scientists and other intellectuals have held views of this kind. Thus, if you think these are reasonable solutions to the problem of reconciling God and evolution, you're in good company. But you're also wrong. That, at any rate, is what I hope to persuade you in Chapter 4.

Chapter 5: God as gap filler

In this chapter, we'll deal with a counterargument that will already have occurred to some readers. Evolutionary theory may account for the apparent design found in the biological world, but there are still many mysteries left to explain – mysteries that may require us to posit the existence of God. Mystery 1: how did life begin? Evolutionary theory can explain the origin of new species from pre-existing species, but it can't explain the origin of life from non-life in the first place. How did the first self-replicating molecules come to exist? Mystery 2: how did the universe begin and why is it so exquisitely suitable for the evolution of life? Mystery 3: how is mind or consciousness possible in a world of mere matter in motion? Each of these questions reveals a gap in the scientific vision of the world, a gap that perhaps only God can fill. As we address them, we'll encounter ideas that to many people might seem quite outlandish, including the idea that this universe is merely one of many, and that Darwinian principles shed light on how universes come to have the properties they do. We'll see, though, that given the current state of play in physics, such ideas are – at the very least – not unreasonable. Also in the realm of surprising conclusions, we'll see that an examination of the evolved mind actually *lowers* our estimate of the likelihood that God exists, rather than raising it. And we'll touch on what I like to think of as the ultimate question: why is there something rather than nothing?

Chapter 6: Darwin and the problem of evil

Evolutionary theory doesn't just eliminate reasons to believe in God; it provides reasons *not* to believe. Have you heard of the problem of evil? It's an argument against the existence of God, the gist of which is that an all-powerful and all-good God would never allow as much 'evil' (i.e., suffering) to exist in the world as we actually find, and thus that there probably is no God. Evolutionary theory radically exacer-bates the problem of evil. The evolutionary process that gave us life involved grotesque quantities of suffering across vast tracts of time. Darwin himself described it as 'clumsy, wasteful, blundering, low, and horribly cruel'.[10] Why would God choose to create life in such a reprehensible manner? Our conclusion in this chapter will be that, if there really were a God, the Creationists would be right – but they're

---

[10] Cited in F. Darwin and Seward (1903), p. 94.

not so there probably isn't. As we make our way to this conclusion, we'll deal with various related issues, such as: are other animals conscious? Do they experience pleasure and pain, or are they merely unconscious biological machines? Even if there really is a God, should we obey him? (You'll need to make sure you read the footnotes if you're interested in the latter question.)

Chapter 7: Wrapping up religion

Chapter 7 deals with a common response to the types of arguments discussed thus far, namely, that they only deal with a traditional, anthropomorphic conception of God. Many claim that the God they believe in is something far grander and more refined than this traditional conception; it is Ultimate Reality, or the Ground of All Being, or the condition for the existence of anything, etc. Maybe such a vision of God is immune to the universal acid of Darwinism. Then again, maybe it's not. This is the subject matter of Chapter 7, the final chapter of Part I.

Part II: Life after Darwin[11]

Moving right along, our second major topic is *philosophical anthropology*, the sub-discipline of philosophy that deals with questions about human beings, their status in nature, and the meaning and purpose of human life. Here are some of the questions we'll be asking in this section (Chapters 8–10):

Does the mind survive the death of the body? • Is the universe conscious? • Are we superior to other animals? • Are we *inferior* to other animals? • Does natural selection inevitably produce progress or is this a misunderstanding of Darwin's theory? • What is the meaning of life? • Does evolutionary theory imply that life is ultimately meaningless?

[11] Parts of this section are based on Stewart-Williams (2002, 2004a).

Chapter 8: Human beings and their place in the universe

This chapter looks at the place of human beings and the human mind in nature, and is, in my opinion, one of the best chapters in the book. We'll see that evolutionary theory poses a challenge to some of our most deep-seated common-sense ideas about the world, including such ideas as that everything in the history of the universe either has a mind or does not, can be classified as either a human being or not, and can be classified as either living or non-living. We'll also see that, from an evolutionary perspective, the mind does not stand outside nature but is a tiny fragment of nature – a fact that has some important implications. One is that our consciousness is not simply consciousness *of* the universe; our consciousness is a *part* of the universe, and thus the universe itself is partially conscious. The section dealing with this topic (entitled 'The conscious universe') probably constitutes my favourite four paragraphs of the entire book; if you read nothing else, I recommend that you read this. Elsewhere in the chapter, we'll see that, although the traditional distinction between humans and animals is still workable after Darwin, it is revealed as no less arbitrary than the distinction between, say, dolphins and non-dolphins; we'll come to the apparently paradoxical conclusion that, although human beings, living things, and minds have not existed forever, there was no first human being, no first living thing, and no first mind; and we'll discover why medical doctors are merely glorified veterinarians and surgeons merely glorified mechanics.

Chapter 9: The status of human beings among the animals

This chapter picks up where the last left off and represents, I hope, one of the main contributions of this book to the field. Many commentators have observed that, like the Copernican revolution before it, the

Darwinian revolution deflated human beings' view of their own importance in the grand scheme of things. As the zoologist Desmond Morris
put it, we are no longer fallen angels but risen apes. But many who
accept Darwin's theory find consolation in the idea that humans are
superior among the animals. Often this is based on the idea that
evolution is progressive – that it involves an inevitable march towards
greater intelligence and complexity, and that we are at the forefront of
this trend. You yourself may hold such a view, and thus you may be
surprised to learn that, in Chapter 9, I will argue that it is wrong in
every particular. My aim is that, by the end of this chapter, you will
cringe every time you hear someone refer to chimpanzees or other
animals as 'sub-human animals' or 'infra-humans', just as you would
cringe if someone described you as a sub-chimpanzee animal or an
infra-turkey. I also hope to persuade you that phrases like 'more
evolved' and 'less evolved' should be jettisoned from your repertoire
of ideas. Along the way, we'll see why the rock star Jim Morrison was an
evolutionary success story even though he died at the age of twenty-
seven; why, from a certain point of view, we are not significantly
smarter than blue whales or even ants; and why it is incorrect to say
that humans are descended from the lower animals. Finally, we'll ask a
question that few ever ask: would the universe be a better place if we
had never evolved?

Chapter 10: Meaning of life, RIP?

In Chapter 10, we deal with one of the biggest questions of them all:
what is the meaning of life? Evolutionary theory could inform our
answers to this question in a number of ways. A common intuition is
that the purpose of an object derives from the intentions of its maker.
If I made a screwdriver, for example, I would have bequeathed it a
purpose: to drive screws. Likewise, if we were made by God, our purpose
would be the purpose for which God created us. What happens, then, if
it turns out that we were *not* created directly by God, but were created

instead by natural selection? There are at least two responses to this question. The first is to substitute natural selection for God: 'If our creator was natural selection, then our purpose in life is whatever natural selection "had in mind" – i.e., passing on our genes.' The second response is very different: 'If our creator was natural selection, then we didn't really *have* a creator at all, and thus our lives have *no* purpose.' It's not much of a choice, but I'm going to argue that one of these options is indeed the case. You'll have to read Chapter 10 to find out which.

Part III: Morality stripped of superstition[12]

Next we'll turn to the important field of ethics and morality. Here are some of the questions we'll be addressing in this section (Chapters 11–14):

> If a behaviour has an evolutionary origin, does this imply that it is natural and therefore morally acceptable? • Does an evolutionary approach to psychology justify the status quo, imply that we can't hold people responsible for the things they do, or justify inequality and war? • Are suicide or voluntary euthanasia ever morally permissible? • How should we treat non-human animals? • Are there ever circumstances in which the lives of human beings should be sacrificed for the good of other animals? • Does exposure to evolutionary theory make people immoral? • Does evolutionary theory imply that, ultimately, nothing is morally right or wrong?

Chapter 11: Evolving good

In the first chapter of Part III, we'll ask whether human morality is an adaptation, crafted by the hidden hand of natural selection. This turns out to be quite a tricky question. On the one hand, there's little doubt that evolutionary theory can shed light on the origins of some of the behaviours that fall within the rubric of morality, including altruism

[12] Parts of this section are based on Stewart-Williams (2004b).

and our characteristic attitudes about certain kinds of sexual behaviour. On the other hand, the morality-as-adaptation hypothesis faces some serious challenges. If morality were a direct product of evolution, why would people constantly argue about what's right and wrong? Why would we have to teach our children to be good? Why would we experience inner conflict between what we think is morally right and what we really want to do? You might expect that, as an evolutionary psychologist, I'd have snappy comebacks for each of these questions. But I don't; I think they represent important criticisms and I don't think that morality is a direct product of evolution. What I do think, and what I'll argue in this chapter, is that morality is a social institution; to some extent, it embodies and reflects our evolved inclinations, but to some extent it also *counteracts* them.

Chapter 12: Remaking morality

In Chapter 12, we'll deal with some questions that have cost a lot of people a lot of sleep over the years. Does evolutionary theory imply that we should adopt the survival of the fittest as an ethical maxim? Does it justify the status quo and the disadvantaged position of women in society? Does it imply that men cannot be held accountable for infidelity or rape? Does it imply that social welfare should be abolished and that dog-eat-dog capitalism is the only acceptable political system? Does it imply that we should forcibly prevent the least fit among us from having children? Does it justify the Nazis' attempt to cleanse the gene pool? More generally, if something we consider bad (e.g., aggression or sexism or racism) can be traced to evolved aspects of the mind, does this imply that it is actually good?

I won't keep you in suspense; you'll be relieved to hear that the answer to all these questions is an unequivocal 'no'. But this might not be for the reason you think. If you know a little about the area already, you will have heard of the 'naturalistic fallacy'. This is generally understood as the fallacy of inferring that because something is natural, it must therefore be good, or that the way things are is the way things ought to be. Technically, such inferences would indeed be logically fallacious; thus, the fact that something has an evolutionary origin does not automatically imply that it is obligatory or even morally acceptable. However, we'll see in Chapter 12 that there is in fact no logical barrier preventing facts about evolution from informing our ethical conclusions, as long as we attend to certain logical niceties. Thus, if there's anything wrong with the ethical conclusions alluded to in the above paragraph (which there certainly is), it is not that they commit the naturalistic fallacy. All will be revealed in Chapter 12.

Chapter 13: Uprooting the doctrine of human dignity

This is another candidate for my favourite chapter of the book. In it, we'll survey the implications of evolutionary theory for a series of topics in applied ethics, including suicide, euthanasia, and the proper treatment of non-human animals. The thread knitting the chapter together is an important trend in moral thinking known as *the doctrine of human dignity*. This refers to the view that human life is infinitely valuable, whereas the lives of non-human animals have little value or even none at all. We'll see that the universal acid of Darwinism dissolves this ancient dogma, and that this in turn leads us to some unsettling conclusions. It suggests, for example, that the notion that human life is supremely valuable is a mere superstition; that there may be circumstances in which it is morally acceptable to take an innocent life; and that there may be circumstances in which it is completely *im*moral and unethical to keep a person alive.

The demise of the doctrine of human dignity also has important implications regarding the treatment of non-human animals. Our species has a long track record of treating other animals poorly; some even liken our treatment of the animals to the Nazi Holocaust. While we've struggled over recent centuries with moral issues such as slavery and the rights of women, few have thought to question the morality of the way we treat members of other species. Recently, however, aware-ness of the issue has grown, and there is now even a word for discrim-ination on the basis of species membership: *speciesism*. In Chapter 13, we'll see that a Darwinian perspective supports the view that speciesism is just as morally objectionable as other forms of discrimination, such as sexism and racism – after all, a universe with less suffering is better than one with more, and, with the doctrine of human dignity safely out of the picture, it makes no moral difference whether the suffering indi-vidual is a human being or some other creature. The Australian philosopher Peter Singer has gone as far as to argue that the animal liberation movement is the single most important liberation move-ment in the world today, more important even than the social move-ments combating sexism and racism.[13] He has also argued that in certain circumstances, the life of a chimpanzee or a pig may be worth more than the life of a human. A moral system anchored in evolu-tionary theory is entirely consistent with Singer's views. I suspect that some readers will see this as a radical and unreasonable position, and dismiss it out of hand. It would be interesting to know what they'll think by the time they get to the end of Chapter 13.

Chapter 14: Evolution and the death of right and wrong

In the final chapter of the book, we consider two main questions. The first is whether exposure to evolutionary theory makes people bad. Many Creationist critics of Darwin claim that his theory purges existence of any ultimate meaning and reduces the value of human life to zero. In doing so, it destroys morality, making people selfish, sexually promiscuous, and violent (so they argue). This concern is exemplified by the famous (possibly apocryphal) words of a

---

[13] Singer (2002b).

nineteenth-century bishop's wife: 'Let us hope that what Mr Darwin says is not true; but, if it is true, let us hope that it will not become generally known.' Many hold this view. But do we really have to deny our evolutionary origins and hold false and groundless beliefs about the universe in order to be good, treat one another nicely, and care about one another's welfare? Could it be that it's actually safer and more effective *not* to tie morality to religion? That's the first topic for Chapter 14.

The second relates to the question of whether our moral beliefs – *any* moral beliefs – are objectively true. You probably feel pretty confident about the veracity of your moral convictions. You feel that murder is wrong and helping people is right, and that anyone who thinks otherwise is simply incorrect. I feel the same way. But think about this: if we were intelligent ants, we would think individual rights were an evil; if we were termites, we would have no moral qualms about reproductive love among siblings (indeed, we'd favour it); and if we were bees, we would consider it a sacred duty to kill our nest mates. We would have these attitudes because they would fit us, as ants, termites, or bees, to the evolved lifestyle of our respective species. But that's precisely the reason we have the moral attitudes we actually do have: because they fit us to the evolved lifestyle of the species *Homo sapiens*. Our moral beliefs are informed by desires and emotions that are there solely because they helped our ancestors pass on their genes. How, then, do we know that our moral beliefs are objectively true? More to the point, how do we know that *any* moral beliefs are objectively true? The verdict of Chapter 14 is that we *don't* know that they're true, and that in fact they're not. In the final analysis, there is no such thing as right or wrong. This does not imply, however, that we can or should dispense with morality. On the contrary, in the last pages of the book, I argue that evolutionary theory helps make the case for a utilitarian approach to ethics – that is, an approach that judges the rightness or wrongness of our actions in terms of the effects of those actions on all involved.

Conclusion

So, that's our subject matter: God, man, and morality in the light of evolution. These topics are densely interconnected, such that whenever evolutionary theory has implications for one, this usually has downstream implications for the others. For example, evolutionary theory lowers the probability that we are the privileged creation of a supernatural being. This has implications concerning our place in nature (it lowers the probability that non-human animals exist merely to satisfy our needs), which in turn has implications within the domain of ethics (it forces us to re-evaluate the way we treat other animals). This example also illustrates something about the structure of the book, namely, that Part I (on the existence of God) lays the groundwork for much of what follows, and that Parts II and III (on humankind and ethics, respectively) contain the more original and less widely discussed ideas. Regardless of your starting point, though, I hope you'll take the entire journey with me. Along the way, we'll meet a diverse and colourful cast of characters, including Creationists and Theistic Evolutionists, vitalists and mechanists, evolutionary psychologists and Blank-Slate Darwinians, Social Darwinists and human supremacists. We'll challenge some commonly held beliefs, such as that Pope John Paul II accepted evolutionary theory; we'll assess some ancient dating advice (from Plato, no less); and we'll even contemplate the ultimate destiny of the universe. Last but not least, we'll reach definitive answers to some age-old mysteries: which came first, the chicken or the egg? Why do innocent people suffer? What happens when we die? There's a lot to get through, so we'd better make a start. The first item on the agenda is God.

PART I

Darwin gets religion

TWO

Clash of the Titans

All flesh is not the same flesh: but there is one kind of flesh of men,
another flesh of beasts, another of fishes, and another of birds.

1 Corinthians 15:39

The evolutionary hypothesis carried to its logical conclusion disputes
every vital truth of the Bible. Its tendency, natural if not inevitable, is
to lead those who really accept it, first to agnosticism and then to
atheism. Evolutionists attack the truth of the Bible, not openly at first,
but by using weasel words like 'poetical', 'symbolical', and 'allegorical'
to suck the meaning out of the inspired record of man's creation.

William Jennings Bryan (1925/2008)

Theology made no provision for *evolution*. The biblical authors had
missed the most important revelation of all! Could it be that they were
not really privy to the thoughts of God?

Edward O. Wilson (1998), p. 6

What in God's name are we talking about?

There is no longer any reasonable doubt about whether evolution
happened. It did, and anyone who's interested in gaining an accurate
picture of the world needs to adjust themselves to this fact. Unfortunately, this is often easier said than done, especially when the

21

adjustments need to be made in sensitive areas. If you've ever challenged someone about their religious beliefs, you'll know that religion is just about the most sensitive area there is. Evolutionary theory raises difficult questions for believers, encroaching on territory that was traditionally the province of religion. That solitary fact almost singlehandedly explains why the theory is so controversial even today, more than 150 years after it was first put forward. And although there's no longer any reasonable debate about evolution itself, there is still reasonable debate about the implications of evolutionary theory for the big questions of life. This includes the question of God's existence. The purpose of this section of the book (Chapters 2 to 7) is to examine the implications of Darwin's theory for theistic belief.

But before getting started, we need to make sure we know what we're talking about. The concept of God has been among the most influential concepts in the history of the human species. But what exactly is God? Without defining the term, we're not in a position to be able to say whether we think that he (or she or it) exists. The problem is, though, that trying to pin down a definition of God is a little like trying to nail jelly to a tree. The definitions on offer vary so widely – not only between religions but also within any given religion – that when two people claim to believe in God, they often turn out not to believe the same thing at all. In fact, sometimes they differ as much from one another in what they believe as each differs from the atheist.

For some, God is literally a Big Guy in the Sky; for others, he is a disembodied and genderless mind. For some, God is a jealous and angry tyrant; for others, he is a loving and forgiving Father. For some, God is a personal being; for others, he is an impersonal force or principle. For some, God directs events and answers prayers (and if sometimes he doesn't, he always has a good reason for this); for others, God takes a hands-off approach, and rarely or never intervenes in the orderly running of the universe. For some, God acts in real time within people's lives and in the course of history; for others, he exists outside time and space. For some, God is something separate from nature; for others, he is identical with nature or the laws of nature. Some try to avoid these contradictions by claiming that God is somehow all these things and more, or that the divine transcends the laws of logic and embraces paradoxes – but that itself is another conception of God, competing with the non-paradoxical ones. To be fair, theologians do tend to agree on a basic wish list of traits for their God: he is perfect, all-loving, all-good, omniscient, omnipresent, and omnipotent.[1] However, the traits emphasized by theologians tend not to occupy centre stage for the rank-and-file of religious believers, whose God is primarily about answering prayers, administering justice, and guaranteeing life after death. There is no getting away from the fact that there is little agreement about what exactly God is. And we've only just scratched the surface. There are numerous other conceptions and descriptions of the divine; here's a sampling of some of my favourites:

God is: the Infinite • The Ultimate • The Absolute • The Absolute Unknowable • Brahman (Supreme Cosmic Spirit) • The Unmoved Mover • The First Cause • The cause of all causes • The explanation for all explanations • The condition for the existence of anything • Self-Transcending Fecundity • A detached and transcendent demiurge • Thought thinking on itself • A circle whose centre is everywhere and whose circumference is nowhere • Mind on a scale beyond human comprehension • The supreme rational agent • Maximally excellent being • The *ens realissimum* (the most real being) • A being greater than which none can be conceived • A necessary being that has its necessity of itself • A being whose essence is identical with its existence • The object of ultimate concern • The ultimate Ground of All Being • A depth at the centre of life • The *élan vital* • The Eternal Thou.

Unfortunately, the vast array of God concepts makes it very difficult to determine whether God and evolution can comfortably coexist in a logically consistent view of the world. Which God are we talking about? In this and the next few chapters, I deal first and foremost with the most

[1] Worthy questions remain, such as whether an omnipotent God would have the power to do evil, or to do something imperfectly, or to cause himself not to exist. On the God of theology, see, e.g., Swinburne (1993).

widespread conception of God: God as a supreme intelligence that deliberately created the universe and who continues to act in the world today. This strategy is likely to get some readers' hackles up. Sophisticated religionists may complain that religious people no longer believe in the childishly anthropomorphic God of earlier ages. They may object that my approach doesn't give the religious perspective a fair trial, and that to attack the simpler conceptions of God with heavy-duty arguments and evidence is a little like donning a full suit of armour to attack a hot fudge sundae (to borrow a line from Kurt Vonnegut). I ask for these readers' patience. Although I start with the simpler conceptions, I turn my attention later to some of the more sophisticated ones. In Chapters 4 and 5, I consider the God of deism; in Chapters 5 and 6, I consider the God of traditional theology; and in Chapter 7, I consider some more abstruse conceptions of God, such as the notion of God as a depth at the centre of life. In this chapter, though, we'll deal with the God of the Creationists: that subgroup of believers who reject evolutionary theory and adhere to a literal interpretation of Genesis.

Taking the Bible at its word

Some of the biggest questions faced by human minds are questions of origin. All cultures have myths and theories concerning the origin of the earth, the origin of the heavens, and the origin of life. Of these, the most popular tend to be those focusing on the latter issue: how did we and the other animals come to exist? In the early twenty-first century, there is really only one scientific answer to this question: we evolved. In popular culture, though, there is another contender. This is the position known as *creationism*. Creationism is rooted in a literal reading of Genesis, the first book of the Christian Old Testament (which is also the first book of the Hebrew Bible and the inspiration for the Islamic creation story found in the Koran). Creationists hold that each species was created separately and independently by God, and is a static and unchanging entity. In their view, human beings are the children of

Adam and Eve, not of monkeys. Creationists reject evolutionary theory as a huge mistake or even a Great Lie, invented (perhaps by the Devil) to lure people away from God and spread atheism. (Apparently the conspiracy has been only moderately successful, for, as we'll see in later chapters, many people who believe in evolution also believe in God.)

Creationists come in various different stripes. Most important among their ranks are the Young Earth Creationists (YECs) and the Old Earth Creationists (OECs). YECs believe that God created the earth and all life during six twenty-four-hour days in the geologically recent past. This is usually assumed to have been between 6,000 and 10,000 years ago. Some YECs are quite specific about the date of the creation; according to one famous estimate, made by Bishop Ussher in the 1650s and based on the genealogies in the Bible, the earth was created in 4004 BC, on 22 October. According to a number of YECs, Eden before the Fall was a peaceful vegetarian paradise. It was not until the fruit-related transgressions of Adam and Eve that animals started killing and eating each other and that death entered the world. Most YECs also believe that around 4,400 years ago there was a great flood which engulfed the entire planet and wiped out many of the life forms around at that time. This includes the dinosaurs.

So that's what YECs believe. OECs, on the other hand, recognize that the earth is very much older than 6,000 or 10,000 years. Some even accept the scientific estimate of the earth's age as 4.6 billion years, and accept that different plants and animals occupied the earth during different epochs. Nonetheless, they still hold that each species was created separately by God. Many OECs make the interesting suggestion that God engaged not in a single bout of creation, but in several: for instance, he created the dinosaurs around 250 million years ago, and then created humans and other animals we know and love more recently. Intellectually, Old Earth creationism is a vast improvement on its Young Earth counterpart. After all, long before Darwin put forward his theory, geologists had demonstrated that the earth was much older than 6,000 years old, and the rapidly growing fossil record showed that different groups of species had populated the planet during different periods in its long history. Old Earth creationism takes this into account, and thus is at least consistent with pre-Darwinian science. But although this makes it superior to Young Earth creationism, it still faces major challenges and difficulties, as we'll soon see.

Darwin's dangerous idea

Young Earth or Old, creationism provides one answer to the question of how we came to exist: God created us, directly and immediately. The other answer, of course, is that we evolved. According to evolutionary theory, and contrary to the view of the Creationists, species are not static; they change and evolve over time. Every species we see today evolved from an earlier species, which itself evolved from an earlier species, and so on and so on, ultimately tracing back to the origin of life on earth 3.8 or so billion years ago. All life on earth is genealogically related, and the main mechanism behind evolutionary change is natural selection (roughly speaking, the survival of the fittest).

If you ask most people who came up with the idea of evolution, they'll tell you it was Charles Darwin. Not so. The idea had been in the air for a long time before Darwin set pen to paper and wrote his world-shaking, epoch-making, paradigm-breaking book, *On the Origin of Species by Means of Natural Selection*. In the late eighteenth century, Darwin's grandfather, Erasmus Darwin, had speculated about the possibility of evolution,[2] and soon afterwards, the French naturalist Jean-Baptiste Lamarck had made an unsuccessful attempt to identify the causes of evolutionary change. Another, equally unsuccessful attempt came in the form of Robert Chambers' anonymously published *Vestiges of the Natural History of Creation*.[3] But although Darwin

---

[2] His best-known articulation of the idea was the posthumously published poem, *The Temple of Nature* (E. Darwin, 1803), perhaps the only important work in the history of science to be written in rhyming couplets.

[3] Chambers (1844).

did not originate the idea of evolution, before he entered the fray, evolution was viewed as fashionable pseudoscience, similar in stature to phrenology. Darwin turned this around, winning intellectual respectability for the concept and quickly persuading most scientists that it was fact and not fiction.

The turning point in evolution's fortunes was 1859, the year of the publication of the *Origin of Species*. Darwin's book had two main aims. The first was simply to argue that evolution had taken place – that is, to argue for what I'll call the *fact* of evolution. Darwin made a forceful case for evolution, and we'll consider some of the evidence for it soon. The second aim was, if anything, more important and ambitious; it was to explain *how* evolution occurs – that is, to put forward a *theory* of evolution to explain the fact of evolution. The mechanism Darwin postulated was natural selection. (Note that natural selection was also independently discovered by the naturalist Alfred Wallace, and hinted at by several others.) This is what the philosopher Daniel Dennett said about the theory: 'Let me lay my cards on the table. If I were to give an award for the single best idea anyone has ever had, I'd give it to Darwin, ahead of Newton and Einstein and everyone else.'[4] Strong praise! But it captures a widespread sentiment. A lot of people just fall in love with Darwin's theory.

Let's see what all the fuss is about. Natural selection has three main elements: differential reproduction; variation; and inheritance. In plain English, these boil down to the following statements: (1) some individuals have more offspring than others; (2) individuals differ from one another; and (3) offspring tend to resemble their parents. When you put these three elements together, you automatically get evolutionary change. It works like this. The first plank in the theory is differential reproduction. Every type of organism produces more offspring than can possibly survive. Left unchecked, populations would increase geometrically (2, 4, 8, 16, etc.). This doesn't happen

4 Dennett (1995), p. 21.

though, which implies that many organisms die without reproducing. As Darwin put it, there is a 'struggle for existence' at every stage. What determines which individuals prevail in this struggle? It's partly a matter of chance: one animal is struck by a meteorite while another narrowly escapes (survival of the luckiest, as Stephen Jay Gould put it). But it's not completely random, and this brings us to the second element in Darwin's theory: variation. No two individuals are exactly alike. Some are taller, some are faster, some are stronger than their neighbours. And in some cases, quite by chance, individuals possess traits that make them more likely to survive and reproduce. For example, some individuals are slightly better than their companions at obtaining food, or escaping predators, or caring for their offspring. All else being equal, these individuals will leave behind more descendants than their neighbours.

The third and final ingredient in Darwin's theory is inheritance. This refers to the fact that parents tend to pass on their own attributes to their offspring. If a trait that increases an individual's likelihood of surviving and reproducing is heritable, that individual's offspring will be more likely than chance to possess the same trait. Because the individual has more offspring, and because the offspring are likely to inherit the trait that gave it this reproductive advantage, the trait will be more common in the next generation. If this continues over many generations, the trait will eventually displace other variants. It will, in other words, be selected.

An example will make things clearer. Let's imagine a small population of squirrels. To keep things simple, we'll imagine just three, and we'll assume that one of these is slightly faster than the others. Thus we have two slow squirrels and one fast one. First question: why is the fast squirrel fast? Well it's *not* because being fast is useful; he just happens to have a random genetic mutation that makes him faster.[5]

---

[5] Note that Darwin didn't know anything about genes or mutations; the integration of his idea of natural selection with genetics did not take place until the early-to-mid twentieth century.

But although he's not faster *because* it's useful, it just so happens that being faster *is* useful, because it means that he's more likely to escape from predators. Because he's more likely to escape from predators, he's more likely to survive, and because he's more likely to survive, he's likely to have more offspring than his slower counterparts. Now let's imagine that the first slow squirrel has one offspring, which inherits his slowness. The second unfortunately has none; she gets eaten before she gets the chance. However, the fast squirrel, because he's fast, lives longer and manages to have *two* offspring, both of whom inherit his fastness. Thus, in the second generation, there is one slow squirrel and two fast ones. Recall that, in the first generation, it was the other way around: there was one fast squirrel and two slow. So, in the first generation, fast squirrels occupied one-third of the population, but now, in the second, they occupy *two*-thirds. The trait of fastness is being selected. That's all natural selection is: heritable traits that increase the chances that an organism will reproduce automatically increase in frequency in the population. If this process continues for long enough in our squirrel population, the fast squirrels will crowd out the slow ones. Squirrels will have evolved to be faster animals.

Natural selection is a very simple idea, at least in its broad outline. But it's also an incredibly powerful idea. The true power of natural selection becomes evident when you add just one more ingredient: time. Over incomprehensibly vast stretches of time, the slow accumulation of serendipitously advantageous traits really adds up. It can result in the formation of complex adaptations (such as eyes and wings and the elephant's trunk and the human brain). And ultimately it can result in the genesis of entirely new species. Part of the magic of Darwin's theory is that all the complexity in the biological world is explained in terms of a principle so straightforward that it can be grasped by an intelligent ten-year-old. No one grasped it, however, until Darwin came along and set his mind to the task. In doing so, he achieved the status of one of the greatest scientists who ever lived.

Why evolution is true

Darwin's theory is clearly inconsistent with the Creationist account of the origin of species. For one thing, natural selection would have needed much longer than 6,000 years to produce the abundance of flora and fauna we see around us today. This contradicts the YECs' estimate of the age of the earth. The idea that, prior to the Fall, the earth was a death-free vegetarian paradise is also hopelessly inconsistent with evolutionary theory (not to mention common sense). Animals were killing and eating each another for millions of years before Eve tasted the forbidden fruit. More importantly, though, evolutionary theory states that current species evolved from earlier ones, and that any two species share a common ancestor if you trace it back far enough on the family tree of life on earth. This directly contradicts the Creationist claim that each species was created separately in its present form. Given the short span of human life and recorded history, it's only natural for people to have assumed that species were static and unchanging, just as our limited perspective makes it natural to view the earth as flat and motionless. However, just as scientific investigation overturned the earth-centred view of the universe, so too it has overturned the view that species are immutable. There is no longer any reasonable doubt that life came about through a process of evolution. How can I be so confident? For one reason and one reason only: there is now a mountain of evidence for evolution – evidence that it happened and evidence that its primary cause was natural selection.[6] In this section, we'll look at some of the most interesting and persuasive evidence for evolution.

We'll start with the fossil record. Scientists now have literally millions of fossils, and have pieced together some excellent sequences showing the transmutation of one kind of organism into another.

---

[6] For accessible summaries of the case for evolution, see Coyne (2009); Dawkins (2009); Fairbanks (2007); Miller (1999); Prothero (2007); Theobald (2007).

This includes the celebrated sequence documenting the evolution of the modern horse, starting with the dog-sized *Hyracotherium* (formerly known as *Eohippus*), which lived around sixty million years ago. It also includes the sequence of fossils documenting the evolution of whales from four-legged land mammals. In addition, scientists have found transitional forms between major groups of modern animals. The best known is *Archaeopteryx*, an intermediate between reptiles and modern birds. If you google 'Archaeopteryx,' you'll find images of a strange-looking animal bearing an eerie resemblance to both a dinosaur and a modern bird. It looks like this because it actually *is* a transitional form between these groups of animals. (That's right; for those of you who didn't know already, birds evolved from a species of dinosaur.)[7] Scientists have also found intermediates between reptiles and mammals. Significantly, they have *not* found any intermediates between *birds* and mammals. Evolutionary theory makes good sense of this pattern. According to the theory, birds and mammals both evolved separately from reptiles, rather than the one evolving from the other. An absence of bird–mammal intermediates is therefore exactly what we'd expect to find.

Many people think that the fossil record is the main source of evidence for evolution. But this is wrong, wrong, wrong. Even if there were no fossils, the evidence for evolution would be overwhelming. Another line of evidence comes from the striking and often very surprising anatomical similarities found between different groups of organisms. For example, giraffes, rats and people have the same number of vertebrae, and amphibians, reptiles, mammals and birds all have the same four-limb body plan. I personally think that many of the commonly cited examples of anatomical similarities are not such convincing evidence for evolution as most evolutionary biologists think. In many cases, the Creationist could argue that, although the

---

[7] An aside: many zoologists argue that birds literally *are* dinosaurs – the one group of dinosaurs that didn't go extinct sixty-five million years ago.

similarities between organisms are *consistent* with common descent, they are also consistent with the idea that all organisms were created by the same designer, and that his designs were variations on some favoured themes. Admittedly, this might seem to suggest that the creator lacks imagination, which is not a path that most believers would want to take. No doubt, though, there's a way around this problem, and in many cases anatomical similarities are relatively easy to reconcile with the Creationist worldview.

In some cases, however, they are not, and the idea that God recycles his best ideas falls flat on its face. Bat wings have a very different function from human hands and whale flippers, but these varied appendages all have a similar bone structure. The size and length of the bones differ greatly, but they contain similar numbers of bones in a similar configuration. In contrast, bat wings and *bird* wings have exactly the same function (i.e., enabling flight), but a very *different* bone structure. Evolutionary theory explains this effortlessly: bats are more closely related to – and thus more similar to – other mammals, such as people and whales, than they are to birds. Creationism has no equivalently satisfying explanation. If the recurring themes in nature are due to God recycling his best ideas, why doesn't he use the same design for bird wings and bat wings? Why are bat wings less like bird wings than they are like *whale flippers*?

Another persuasive argument for evolution over creationism involves highlighting not just similarities between organisms but *useless* similarities. Many organisms possess features that serve no function. Blind cave salamanders, for example, have rudimentary, functionless eyes, and kiwis and other flightless birds possess rudimentary, functionless wings. Evolutionary theory makes short work of these otherwise puzzling phenomena. Cave salamanders have functionless eyes because they evolved from lizards that did not live in caves and that therefore needed fully functioning eyes. Kiwis have functionless wings because they evolved from birds that could fly and that therefore needed fully functioning wings. How do Creationists explain these functionless appendages? There's a two-word answer to this question: they don't.

Some of the evidence for evolution is just plain weird. This includes the fact that, during early embryonic development, land-dwelling vertebrates (including humans) have distinct gill slits, like those found in fish. This makes perfect sense on the assumption that all land-dwelling vertebrates evolved from a sea-dwelling species, and that some elements of the early developmental programme remain intact (i.e., they were never selected out). However, it makes no sense at all on the assumption that God created each land-dwelling species from scratch. If we were created in our present form, why would we have gill slits? Another weird line of evidence comes from what are called *atavisms*. Occasionally whales are born with hind legs. The evolutionist's explanation for this is that whales evolved from legged land mammals, and that they still possess the genes involved in the development of hind limbs. Usually these genes are deactivated by other genes; sometimes, however, something goes awry during the developmental process and the genes are expressed. Whales with hind limbs are utterly inexplicable on the assumption that each species was created separately. Why would a sane and sensible God include genes for legs in the whale genome?

In addition to this indirect evidence, evolutionary change has been directly observed, both in nature and in the lab. Animal and plant breeders have produced numerous new strains of life by selectively breeding the most desirable specimens from each generation. Scientists have seen insects evolve resistance to insecticides, and bacteria evolve resistance to antibiotics. The evolutionary biologists Peter and Rosemary Grant have shown how various attributes of the finches in the Galapagos Islands – attributes such as beak-length – change in response to environmental conditions.[8] Creationists are quick to counter that these are all examples of what they call

---

[8] Grant and Grant (2008). See Weiner (1994) for an accessible and enjoyable account of their work (an account that won the Pulitzer Prize).

microevolution: a change in the frequency of genes or traits *within* a species. But none involves *macro*evolution: the evolution of one species into another. No one has ever seen macroevolution take place, they argue. This claim is debatable, but even if it were true, it would hardly be fatal. If natural selection can produce small-scale change in the short term, why could it not produce large-scale change in the long term? Unless a compelling answer can be found, a sensible default assumption would be that it could and does. And let's not forget that all the indirect evidence (the fossil record, etc.) suggests that species do indeed evolve from other species.

In Darwin's time, many thinkers came to accept that evolutionary theory applied to non-human life, but denied that it applied to human beings. Even today, some claim that the rest of the biological world evolved, but that humans are the special creation of God.[9] The evidence, though, is clear and unambiguous: we too are products of evolution.[10] An initial point to make is that if God separately created humans and humans alone, then for some strange reason, he stole a lot of his design ideas from natural selection. Many of the design features found in humans can also be found in other animals (the animals that, unlike us, were created by natural selection). So, that's the first mark against the hypothesis that humans were uniquely created. A second point is that, if humans alone were created from scratch, then God went to great pains to make sure that they would fit within the same classificatory system that encompasses the evolved part of the living world – to make sure, in other words, that we would

---

[9] This is not the only claim of this nature to have been made. The Nazi leader Heinrich Himmler was of the opinion that the Aryan people were not products of the evolutionary process; all the other races evolved, but not the Aryans. The only real difference between Himmler's insane claim and the initial one is that Himmler drew the line along racial boundaries rather than species boundaries. It is telling that claims of human exceptionalism – even if we dismiss them as false – raise virtually none of the horror that Himmler's claims do.

[10] See, e.g., McKee *et al.* (2005).

fit into the categories of primate, mammal, animal, etc. It's not obvious to me why he'd think it important to do this.

Then there are the fossils. According to many Creationists, scientists have been unable to come up with a 'missing link' between humans and (other) apes. But scientists have found numerous missing links. Most people have heard of at least some of them, e.g., Homo erectus and Homo habilis. We now have a detailed and thorough sequence of fossils tracing the evolution of humans from our non-human ancestors. I suppose a tenacious Creationist could argue that these were not transitional forms, but separately created creatures that happen to have gone extinct. Why, though, did God create so many species that *appear* to be intermediates, rather than creating entirely different forms?

Like evolution in general, the case for human evolution is not dependent on the fossils. Another highly persuasive line of evidence is the fact that our bodies contain vestiges of our evolutionary past: organs and features that serve no discernible function but which were inherited from ancestral species. My favourite example is goose bumps. When we're cold or frightened or angry, the hairs on our bodies stand on end. Have you ever wondered why this happens? In other mammals, it serves a definite purpose. In fact, it serves several. For one thing, puffing up the fur increases heat insulation and helps keep an animal warm. In addition, when threatened, it can make an individual look bigger, and therefore lower the chances of a rival or predator attacking. Despite the fact that we humans have lost all but a fine layer of hair on most of our bodies, we still get goose bumps when we're cold or afraid. This clearly doesn't serve the function of heat insulation or protection. Indeed, it probably doesn't serve any function at all. It's an evolutionary vestige. We evolved from creatures with much more hair than us, and for whom the goose bump mechanism was useful. It's no longer useful, but the mechanism remains. That, at any rate, is the evolutionist's story. Do you have a better explanation for goose bumps?

Another example of an evolutionary vestige is the tailbone (or coccyx). This structure is almost certainly functionless; it can be surgically removed without any complications over and above those produced by the surgery. Why, then, if we're the special creation of God, do we have it? No doubt a reason could be invented. But that reason had better be more plausible than the idea that the common ancestor of all the apes (including ourselves) had a tail, and that, somewhere along the line, the tail disappeared – but not entirely. The tailbone is another vestige of our evolutionary history, a 'stamp of our lowly origin' as Darwin put it.

Speaking of tails, did you know that, like other non-aquatic vertebrates, for a period of time in early pregnancy, human embryos have clearly identifiable tails? Furthermore, just as whales are occasionally born with hind limbs, human babies are occasionally born with genuine, working tails. Why is that? Well, it turns out that we've all got the genes for tails lying dormant in our genomes and that, once in a while, these genes are expressed and the child is born with a tail. It's extremely hard for Creationists to explain why, if we were created from scratch in our present form, the human genome would contain genes for tails.

Some of the most persuasive evidence for evolution comes from molecular biology. The key finding here is that all life on earth employs the same genetic code. That is, across all species, the same sequences of DNA code for the same amino acids. This constitutes strong support for the common origin thesis (or, I suppose, the Unimaginative Creator thesis). Furthermore, members of different species share a high proportion of their genes, and more closely related species share more genes than less closely related ones. Humans, for instance, share more than 94 per cent of their genes with our closest living relatives, the chimpanzees. One might argue that this is another instance of the creator reusing his designs. The problem is, though, that we don't share only functional genes with chimps; we share non-functional genes as well. Why would this be the case unless we and the chimpanzees had evolved from a common ancestor that itself possessed these useless genes?

That's just a sampling of the evidence for evolution. And we've yet to consider the strongest reason to accept that all life evolved. Evolution is not established by any one piece of evidence; it is the massive convergence of evidence that renders it utterly unreasonable to doubt that evolution has taken place and is still taking place today. Even if Creationists found a way to explain away each and every piece of evidence for evolution, they would still have to explain why all the evidence happens to line up in such a way as to support Darwin's theory. If it's not because the theory is accurate, then it's a coincidence of astronomical proportions, or it's the largest, most ambitious, and most successful con job in history. The case for evolution is so strong now that, if you still believe in the Creationist God, you really need to ask yourself why the creator has gone to such great efforts to trick us into believing, falsely, that life evolved. Dennett points out that the hope that it will be discovered that species did not evolve after all is about as realistic as the hope that it will be discovered that the earth is flat after all. In fact, he goes as far as to suggest that 'anyone today who doubts that the variety of life on this planet was produced by a process of evolution is simply ignorant'.[11] They are ignorant of the evidence and of the unrivalled explanatory power of one of the most successful and enlightening scientific theories of all time: Darwin's dangerous idea.

Creationism's Trojan Horse

And yet there remains a strong, well-organized, and well-funded anti-evolution movement. Most of the opposition comes from Creationists, and takes the form of two main challenges: (1) that evolutionary theory is factually inaccurate, and (2) that the theory causes moral depravity. We'll deal with the latter issue in Chapter 14. As for the former, Creationists have a well-stocked arsenal of arguments aiming to establish the falsity of Darwin's theory. This includes perennial favourites such as: 'evolution is just a theory'; 'evolution is impossible because it would defy the second law of thermodynamics'; and 'something as wonderfully

[11] Dennett (1995), p. 46.

complex as the human eye could not come about purely by chance'. These criticisms reflect simple misunderstandings of the theory, reasons to read more about it rather than reject it. Recently, though, a more sophisticated attack on evolutionary theory has appeared on the scene in the shape of the Intelligent Design (ID) movement.[12] The movement was founded by Phillip Johnson, a retired professor of law who not only questions the scientific consensus on evolution, but also the 'theory' that HIV causes AIDS. ID consists of two main theses.[13] The first is that evolutionary theory is unable to account for at least some biological phenomena. It is a 'theory in crisis'. The second thesis is that the things that evolutionary theory cannot explain *can* be explained on the supposition of some kind of intelligent designer. ID is officially agnostic about the identity of the designer. It could have been God tinkering with DNA, but it could just as well have been a time-travelling biochemist or an alien species.[14] ID theorists are happy to admit that they think it's God, but that's not part of the ID thesis – not officially at any rate.

Opponents of ID routinely complain that it is creationism in disguise; the movement has been described as 'stealth creationism' and 'creationism in a cheap tuxedo'. Considered strictly as an intellectual position, however, ID is not identical to creationism. Some ID advocates are Creationists, certainly, but others are not. Several of the movement's leading lights are 'Theistic Evolutionists' – that is, they accept that all life on earth descended from a common ancestor, but deny that natural selection provides an adequate explanation for this.[15] Thus, intellectually, ID is distinct from creationism. Politically, however, the ID movement is a continuation of the earlier creationist and creation science

---

[12] Some of the most important works on ID, both for and against, include Behe (1996); Brockman (2006); Dembski (1998); Dembski and Ruse (2004); Forrest and Gross (2004); Johnson (1993); Kitcher (2007); Pennock (1999); Shermer (2006).

[13] Kitcher (2007).

[14] Amusingly, the ID arguments have recently been co-opted by people such as the Raëlians, who argue that humans were created by extraterrestrials. Poetic justice!

[15] E.g., Behe (2007).

movements, the goal of which has always been to make elbow room for God in public schools in the United States. Scientists are less patient with ID than they are with religion in general, because they charge – with good reason – that many ID proponents are deliberately concealing their true agenda, which is not to advance science but to turn the public against evolutionary theory and get ID taught in science classes.[16]

But let's ignore the politics and focus on the arguments. The most famous and intuitively appealing argument associated with the ID movement is Michael Behe's *argument from irreducible complexity*.[17] The basic premise is that certain biological structures are too complex to have evolved through natural selection and must therefore be the handiwork of an intelligent designer. The history of evolutionary theory is littered with confident assertions that this structure or that could not have evolved; early candidates included eyes and wings. These claims have not fared well,[18] and thus this general approach seems like a risky one to take. However, Behe has added two new twists to the argument. First, he has shifted its focus from visible structures to microscopic biochemical structures (Behe is a biochemist). And, second, he has elaborated the concept of irreducible complexity.

According to Behe, a structure is irreducibly complex if removing any one of its parts would mean that the structure could no longer function. Watches and mousetraps are examples of humanmade structures that are irreducibly complex. Behe suggests that the biochemical world is replete with such structures, and insists that irreducibly complex structures cannot evolve through natural selection. His most famous example is the bacterial flagellar motor. A flagellum is a projection that bacteria use to move about. In some bacteria, the flagellum functions like a paddle, flapping from side to side. But in others – the ones that really impress Behe – the flagellum functions like a rotary

[16] A strong case to this effect was made by Forrest and Gross (2004).
[17] Behe (1996).
[18] See the special issue of *Evolution: Education and Outreach*, 1(4), 2008.

motor, spinning round and around. Behe maintains, quite reasonably, that the various parts that make up the flagellar motor could not have come about in one single step through random mutation or sexual recombination, because this is simply too improbable. However, he also denies that the system could have evolved through a gradual sequence of steps, because each and every one of the individual parts would need to be in place before the system could work at all. Any intermediate steps would be non-functional. As such, they could not be selected, and would be weeded out before the next step could be taken. Behe concludes that, like a watch or a mousetrap, the flagellar motor must be a product of deliberate design. The same applies to other biochemical systems, such as the immune system, the metabolic system, and the blood-clotting system.

Behe modestly describes the biochemical evidence for intelligent design in the following glowing terms: 'The result is so unambiguous and so significant that it must be ranked as one of the greatest achievements in the history of science. The discovery rivals those of Newton and Einstein, Lavoisier and Schrödinger, Pasteur and Darwin.'[19]

Is there any chance that Behe is being too hasty in drawing this conclusion? You can guess what my answer is; here are my reasons for it. First, promising evolutionary rationales have been posited for many of the systems that Behe claims are irreducibly complex, includ-ing the flagellar motor, the immune system, and the blood-clotting system.[20] We won't examine any of these in detail; it will suffice to note that virtually no expert in this field sees any reason to doubt that so-called irreducibly complex structures can evolve through natural processes. What might be more useful for non-experts is to consider some general problems with Behe's argument.

---

[19] Behe (1996), pp. 232–233.
[20] On the flagellar motor, see Doolittle and Zhaxybayeva (2007); on the immune system, see Inlay (2002); and on the blood-clotting system, see Doolittle (1997).

The first applies not only to Behe but to most Creationist/ID arguments. The problem is that these arguments rely almost exclusively on pointing out supposed failures of evolutionary theory. Proponents of the arguments seem to assume that, if science cannot answer a question, God must be involved. But why should God win by default? Even if a given phenomenon cannot and never will be explained in evolutionary terms, this does not provide any reason to think that God is the true explanation. Holes in one theory do not constitute support for another. If they did, which theory would they support? Would they support Christian creationism or Islamic creationism? Why not the Hindu view that life has existed in its present state for millions of years or forever? Better yet, why not Lamarck's theory of evolution or some other non-Darwinian naturalistic account, as yet undiscovered? The strategy of criticizing supposed flaws in evolutionary theory only seems reasonable if we maintain a blinkered view of the available options. Christian creationism is one of thousands and thousands of unsupported origin beliefs. If supposed flaws in evolutionary theory support any of these beliefs, they support all of them.

In fact, though, they support none. As we've seen, evolutionary theory is backed by a ton of evidence. No creation myth is supported by any good evidence. How would one or a few shortcomings in an otherwise strongly supported theory justify rejecting that theory in favour of another one that has absolutely no support? Let me put it another way. Evolutionary theory explains much of what we know about the biological world. The fact that there remain some obstinate facts that the theory has yet to explain provides no reason to jettison it in favour of a theory that explains *nothing* about the biological world. Of course, some would argue that the God hypothesis actually explains *everything* about the biological world, and everything about everything else as well.[21] But does it really explain anything? Certainly, it's easy enough to say 'God did it.' However, it's just as

[21] See, e.g., Swinburne (2004); Ward (1996).

easy to say 'Frosty the Snowman did it.' The Frosty hypothesis explains things in precisely as much detail as the God hypothesis, and is therefore precisely as good an explanation. In other words, it is *not* a good explanation. A good explanation must have at least enough detail to be consistent with one set of facts but not another. God (or Frosty) could be invoked to 'explain' any possible fact. This tells us that the God hypothesis is vacuous. That in itself doesn't prove that it's false. However, it does undermine any argument that tells us that we must accept God's existence on the grounds that God explains otherwise inexplicable facts about the world. The God hypothesis doesn't explain anything.

The same complaint can be levelled specifically at the ID hypothesis. How was the variety of life on earth created? Imagine if the evolutionist's answer were simply 'natural forces did it'. If that were the whole of the theory, it wouldn't even *count* as a theory. It would barely count as the shadow of a sketch of an outline of a theory. At the most, we could claim that it was a vague hint about the sort of ballpark a promised theory might occupy. But the ID explanation – 'intelligence did it' – is no more detailed than the claim that natural forces did it. How does ID explain the fossil record, or the pattern of differences and similarities among organisms? How exactly did the intelligent designer create the bacterial flagellar motor or the immune system? If it seems that God does a better job of explaining facts that evolutionary theory supposedly cannot, this is only because of an unstated and unwarranted assumption that God can do anything at all. None of the details are ever filled in.

There's one more problem with the such-and-such-could-not-possibly-evolve argument. In essence, any argument of this nature asserts that certain biological structures are amazing, and thus that they must be the products of an intelligent designer. To be fair, no one should deny that the order and complexity found in living things is amazing beyond belief. However, the intelligent designer is presumably even more amazing than the biological structures. If the designer is God, he's *infinitely* more amazing. Thus, positing an intelligent designer does not solve the problem it purports to solve; it makes the problem worse. Once again, this in itself does not *disprove* the existence of God. What it does, though, is take the wind out of the sails of a particular lineage of arguments for God. The whole point of arguments such as Behe's is that we need to posit God to solve a mystery: the mystery of how certain functionally complex structures came to exist. But because the arguments actually substitute this mystery for an even greater one, they don't do what they're intended to do. Thus, any argument of the form 'such-and-such is amazing so God must have done it' must be rejected.[22] Believers should then ask themselves: why *do* I believe in God?

Given the demonstrable weaknesses of the Creationist/ID arguments, we have to ask: why are these arguments so persuasive to so many people? One important reason is that they are consistent with the intuition that complex structures could not come about through simple, mindless processes. Why, though, should we expect our intuitions on such matters to be accurate? The history of science suggests that our intuitions about reality beyond the range of everyday human experience are reliably unreliable. For objects similar in size to us, and for periods of time measured in years or decades, they're usually accurate enough for our purposes. But beyond this all bets are off. Our intuitions systematically fail us on the scale of the very large (the universe described by general relativity), and the very small (the subatomic world described by quantum physics). Much of modern science is deeply counterintuitive, and evolutionary theory is no exception. It's not as mind-bending as relativity theory or quantum physics, but it does require us to leave our untutored intuitions at the door. Unfortunately, though, most people overestimate how useful intuition is as a guide to truth. Intuitive plausibility is the royal road to popular appeal – hence the success of Creationist and ID arguments among non-scientists.

[22] For a related argument, but focusing on complexity/improbability, see Dawkins (2006).

Conclusion

From the moment Darwin put forward his theory, it was obvious to his contemporaries that, if he was right, the biblical account of the origin of life must be wrong. In principle, people might have concluded that the Bible was wrong about the *details* of how God created life, but not about the fact that God existed in the first place. And some did take this route, especially once it became clear that Darwin's theory was here to stay. However, many viewed the biblical account as a package deal, and assumed that, if evolutionary theory challenged one aspect of it, it challenged it all. As one alarmed commentator put it, 'If this hypothesis be true, then is the Bible an unbearable fiction ... then have Christians for nearly two thousand years been duped by a monstrous lie.'[23] Naturally, many of Darwin's contemporaries were highly resistant to accepting the hypothesis. And many people still are. Like the average Victorian in 1859, today's Creationists perceive in Darwin a very real threat to their worldview. But the threat doesn't just come from the clash between evolutionary theory and the biblical account of creation. Indeed, if anything, this is the least of the believer's worries, as we'll see in the next chapter.

[23] Cited in White (1896), p. 93.

THREE

Design after Darwin

The heavens declare the glory of God; and the firmament sheweth his handywork.

Psalms 19:1

For the invisible things of him from the creation of the world are clearly seen, being understood by the things that are made, even his eternal power and Godhead.

Romans 1:20

Natural selection enabled one to see, almost as in a conversion experience, how nature could counterfeit design ... it can be a profound existential experience when one first sees the world not as Paley saw it but through the eyes of Darwin.

John Brooke (2003), pp. 195–6

God the builder

Over the ages, many reasons have been put forward for believing in God, but two have been particularly popular and persuasive with a Christian audience. The first is an appeal to the authority of the Bible; we dealt with this in Chapter 2. The second and more sophisticated reason is *the argument from design*. Stripped to its barest bones, the argument is as follows: certain parts of the world look as though they

45

were designed; design implies a designer; and that designer is God. In this chapter, we'll explore the design argument and its history, and then look at how Darwin's theory rewrites the very terms of the debate.

The argument from design

The argument from design (also known as the *teleological argument*) is one of the most admired and intuitively plausible arguments for the existence of God. It has been in circulation since at least the time of the ancient Greeks, and makes regular appearances in the sacred texts of the world's religions (as we see in the first two quotations at the start of the chapter). Curiously enough, one of the clearest and most beautifully expressed renderings of the design argument came not from a supporter but a critic: the great eighteenth-century philosopher and career sceptic, David Hume. In his posthumously published *Dialogues Concerning Natural Religion*, Hume put the following words in the mouth of his character Cleanthes. (His own views were expressed by the 'careless sceptic' Philo.)

> Look round the world: contemplate the whole and every part of it: you will find it to be nothing but one great machine, subdivided into an infinite number of lesser machines, which again admit of subdivisions to a degree beyond what human senses and faculties can trace and explain. All these various machines, and even their most minute parts, are adjusted to each other with an accuracy which ravishes into admiration all men who have ever contemplated them. The curious adapting of means to ends, throughout all nature, resembles exactly, though it much exceeds, the productions of human contrivance; of human designs, thought, wisdom, and intelligence. Since, therefore, the effects resemble each other, we are led to infer, by all the rules of analogy, that the causes also resemble; and that the Author of Nature is somewhat similar to the mind of man, though possessed of much larger faculties, proportioned to the grandeur of the work which he has executed. By this argument a posteriori, and by this argument alone, do we prove at once the existence of a Deity, and his similarity to human mind and intelligence.[1]

[1] Hume (1777), Part II.

When it comes to apparent design in nature, the most striking examples are found in the biological world, the world of life, which is simply teeming and overflowing with purpose-serving organs and nifty contraptions. Not surprisingly, one of the most influential versions of the design argument drew heavily on observations of the living world. Its author was Archdeacon William Paley. In a famous passage, Paley observed that, if you were crossing a heath and came across a rock, you would never assume that someone had made it. But if you found a *watch* – well, that's a different story. The parts of a watch are perfectly arranged so that it can perform a certain function. If they were not so arranged, the watch could not perform that function. According to Paley, precisely the same argument applies to biological structures. The eye, for example, is an engineering marvel. Surely, then, it must be the work of a marvellous engineer. As Paley put it, 'there is precisely the same proof that the eye was made for vision, as there is that the telescope was made for assisting it'.[2] Likewise, the hand appears to be made for grasping and manipulating objects, and the digestive system for extracting nutrients from food. Presumably, then, these things were designed to fulfil these purposes. Paley summed up the argument like this: 'The marks of *design* are too strong to be gotten over. Design must have a designer. The designer must have been a person. That person is God.'[3] (Note that the last step in the argument – the identification of the designer with God – is often glossed over, but it is a distinct step and requires independent justification.) Darwin stumbled upon the work of Paley when he was studying theology at Cambridge, and later wrote: 'I do not think I hardly ever admired a book more than Paley's "Natural Theology". I could almost formerly have said it by heart.'[4]

Before 1859, science seemed only to strengthen the design argument. It revealed a universe with vastly more order and intricacy than

---

[2] Paley (1802), p. 16.
[3] Paley (1802), p. 229.
[4] Cited in F. Darwin (1887b), p. 219.

anyone had expected to find: the clockwork precision of the orbit of the planets, the minute detail found in biological structures, the mathematical beauty and economy of the physical laws knitting together the universe. In revealing these hidden wonders, science increased people's awe at God's craftsmanship and ingenuity. To admire nature was to admire God's genius. Admittedly, science was simultaneously making it more and more difficult to take the Bible seriously; as mentioned, geologists had shown that the earth was much older than scripture seemed to suggest, and biblical scholarship had shown that the Holy Book was a manmade document, riddled with contradictions and historical inaccuracies. But although science made it harder to believe in the Bible, for many people it made it easier to believe in God. This was the case, at least, until Darwin came along and stole the spotlight.

Hume's critique of the design argument

It is a measure of the persuasiveness of the argument from design that it continued to hold sway with serious thinkers for so long. I say this because its popularity flew in the face of some very serious philosophical concerns. Nearly a century before the publication of the *Origin*, David Hume had exposed some of the weaknesses of the argument. Even assuming that design implies a designer, he noted, why should we assume that the designer was the God of the monotheistic religions? The design argument provides just as good (or just as bad) an argument for Zeus or Thor as it does for the Abrahamic God. And why should we assume that there was just one designer rather than a team? In the creation myths of many cultures, there are two creators: a male and a female. The design argument provides just as strong support for *these* creation myths as it does for the idea that the world was created by one God. We might even wish to argue that the evidence provides *stronger* support for the two-creator hypothesis. When Darwin first encountered the animals of Australia and observed just how very different they were from other animals in other parts of the world, he remarked in his diary: 'An unbeliever in everything beyond his own reason might exclaim "Surely two distinct Creators must have been [at] work."[5] The only reason people accept that the design argument proves the existence of a single God, as they imagine him to be, is that they already believed in this God for other reasons.

Hume raised further difficulties with the design argument. Why should we assume that the designer was all-powerful or perfect? The universe hardly seems perfect, and even if it were, a perfect creation does not necessarily imply a perfect creator. As Hume noted:

> If we survey a ship, what an exalted idea must we form of the ingenuity of the carpenter, who framed so complicated, useful, and beautiful a machine? And what surprise must we feel, when we find him a stupid mechanic, who imitated others, and copied an art, which, through a long succession of ages, after multiplied trials, mistakes, corrections, deliberations, and controversies, had been gradually improving? Many worlds might have been botched and bungled, throughout an eternity, ere this system was struck out: much labour lost: many fruitless trials made: and a slow, but continued improvement carried on during infinite ages in the art of world-making.[6]

That's still not the end of the theist's headaches. How do we know the designer is good rather than evil, God rather than Satan? How do we know the designer still exists? Maybe God is dead! The design argument certainly provides no reason to think otherwise. Another possibility – one that Hume did not mention – is that the designer was a non-conscious automaton. You might ask how a non-conscious automaton capable of designing a universe could have come into existence. And that's a good question – but it's no less a problem explaining how a *conscious* designer so capable could have come into existence!

We see, then, that almost a full century before the publication of the *Origin*, Hume had raised grave doubts about the design argument.

---

[5] January 1836, cited in Keynes (2001), p. 402.

[6] Hume (1777), Part V. Interestingly this suggestion bears a notable resemblance to the idea of Darwinian evolution, and to an idea that we'll consider in Chapter 5: Lee Smolin's theory of the evolution of universes.

However, neither he nor anyone else had been able to think of a better explanation for the apparent design in nature, and thus the argument retained much of its force. Indeed, Hume himself reluctantly concluded that the cause of the universe probably bore some remote resemblance to human intelligence. In 1831, when a young Darwin set forth on his historic voyage aboard the *Beagle*, the idea that the design in nature was evidence for the existence of God was the orthodox view.

But Darwin was soon to change everything.

Unravelling design

Most people would agree that certain parts of the natural world look as though they were designed. Before Darwin, philosophers thought there were two possible explanations for this: either they really were designed or they came about through chance alone. The idea that they came about through chance alone stretches credulity to breaking point, and thus we're left with design (so the argument goes). In hindsight we can see that, even if we accept the design/chance dichotomy, this is a weak argument. After all, no matter how unlikely chance is as an explanation, design may be even less likely. But this is a moot point, for Darwin completely undermined the design/chance dichotomy. He provided a 'third way': an explanation for how a mindless natural process could create the illusion of design. Darwin himself commented on the philosophical implications of his third way in an essay he wrote on the subject of religion (an essay which he never intended to be published but which surfaced after his death):

> The old argument of design in nature, as given by Paley, which formerly seemed to me so conclusive, fails, now that the law of natural selection has been discovered ... There seems to be no more design in the variability of organic beings and in the action of natural selection, than in the course which the wind blows.[7]

[7] Darwin (2002), p. 50.

Paley had written that 'the examination of the eye was a cure for atheism'.[8] Darwin removed this particular remedy from the market. For many, this was quite a shock to the system. Before Darwin, science had only seemed to add weight to the design argument. Now a scientific theory was spiriting away the jewel in the crown of the argument: design in the living world. Darwin's theory provided an alternative way to account for this design. And this raised an uncomfortable possibility: if God is not needed to explain the design in nature (which was generally considered the best evidence for a designer), then maybe God does not exist. By eliminating the need for any supernatural theory of the origin of species or the functionality of adaptations, Darwin did what Hume had been unable to: he made it psychologically possible for people to let go of the design argument and thus to let go of God. In other words, in this area of philosophy, Darwin had a greater impact than one of the greatest philosophers ever: David Hume.

How might the theist respond to this? One response would be to argue that although evolutionary theory does drastically weaken one of the most potent arguments for the existence of God, this in itself does not demonstrate God's *non*-existence. Plenty of believers were and are oblivious to the design argument, and would be unlikely to agree that the elimination of an argument they've never heard of compels them to give up the ghost, as it were, and recant their belief in God. Among those who *have* encountered the argument, the majority believed already for other reasons. At the most, the design argument bolstered their faith, perhaps easing the nagging doubt that they had no reasonable reason to believe. Individuals in this boat are unlikely to be concerned about the demolition of the argument at the hands of Darwin. Furthermore, some Christian thinkers rejected the design argument even before Darwin pulled the rug out from under its feet. Why, then, should the argument's demise be of any concern to those who believe in God?

[8] Paley (1802), p. 23.

This is a fair question. However, it should be noted, first, that even among those who initially believe in God for other reasons, the design argument often strengthens their belief. As such, as soon as they recognize that the argument fails, their belief should be weakened accordingly, even if it's not dropped altogether. It would be intellec-tually dishonest to continue believing in God with just as much gusto as ever, but to switch to the claim that God does not require proof, or that one's belief had been based on faith or other grounds all along. Anyone employing rational arguments to support the belief in God should be willing to give up this belief (or at least to reduce the strength of their conviction), if and when those arguments are shown to be wanting.

Second, in pointing out that the design argument is not the initial reason for belief, one raises the question of what the initial reason really was. For most people, the reason they hold the religious beliefs they do is that when they were young – too young to think critically – they were spoon-fed these beliefs by their parents and others they assumed were trustworthy authorities.[9] But this cannot be a reliable means of forming accurate beliefs. Different authorities say different things, and they can't all be right. Darwin's assault on the design argument brings these issues to the fore. It therefore represents an important challenge to theistic belief, even for those who believe for other reasons. Whichever way you cut it, God is on shakier ground after Darwin than he was before.

Conclusion

We've seen that evolutionary theory attacks traditional religious belief on two main fronts: it implies that a literal reading of Genesis is false, and it steals the thunder from the argument from design. It is

[9] Imagine how many fewer people would be believers if no one were introduced to religion until they were, say, twenty years old. Most religious people would concede that, in such circumstances, there would be far fewer believers in the world. They therefore tacitly accept that religious beliefs sound implausible.

hardly surprising, then, that when Darwin first unveiled his theory, it provoked a firestorm of controversy, catalysing a battle between science and religion equal in stature to Galileo's famous conflict with the church. Many viewed evolutionary theory as utterly incompatible with theistic belief. However, the concept of God has proven to have a long shelf life. It has resisted extinction by continually adapting to the new intellectual environment in which it finds itself. Lots of people now maintain a belief in God alongside an acceptance of the central tenets of the scientific worldview, including the idea that life evolved through natural selection. In the next chapter, we'll look at how they do it and whether their efforts stand up to scrutiny.

FOUR

Darwin's God

Extinguished theologians lie about the cradle of every science, as the strangled snakes besides that of Hercules.

> Thomas Huxley, cited in Browne (2006), pp. 94–5

This century will be called Darwin's century. He was one of the greatest men who ever touched this globe. He has explained more of the phenomena of life than all of the religious teachers. Write the name of Charles Darwin on the one hand and the name of every theologian who ever lived on the other, and from that name has come more light to the world than from all of those.

> Robert Green Ingersoll (1884), from *Orthodoxy*, cited in Ingersoll (2007), p. 168

> Some call it Evolution,
> And others call it God.
>> Williams Herbert Carruth (1909), p. 2

Reconciling God and Darwin

Charles Darwin was absolutely terrified of going public with his theory of evolution. The basic idea of natural selection first dawned on him in 1838, but it was twenty years before he finally announced it to the world. In other words, for two whole decades, he sat on one of the most important scientific theories of all time, only sharing his bold speculations with a handful of close friends. And throughout that time, he worried and he worried about how the public at large would respond when he finally unveiled the theory. In 1844, still fourteen years before that fateful day arrived, Darwin admitted in a letter to a friend that he was 'almost convinced' that species change over time, and described this admission as 'like confessing a murder'.[1] He reported having nightmares about being beheaded or hanged for his ideas. No one knows for sure whether this is the main reason he held back his theory for so long; some argue he was just making sure he had a strong enough case to support such a radical new idea. Either way, there is no doubt that Darwin was deeply concerned about the uproar his theory would provoke.

But despite the years of worry and the sleepless nights, when Darwin finally did come out of the closet and confess his murder, he wasn't hanged or beheaded for it. The public reaction to his theory was surprisingly mild. The idea was certainly controversial, but it was a lot less controversial than Darwin had anticipated and than common wisdom would have us believe.[2] From the moment the *Origin* hit the bookstands, there were Christians who accepted evolutionary theory. Some even welcomed it with open arms; one commentator called it 'an advance in our theological thinking'.[3] The feeling in many quarters was that if scientific investigation reveals that Genesis cannot be taken literally, and that the varied species on this planet emerged through an evolutionary process, then surely we now know more about God and his creation than we did before, and our understanding of the divine has been strengthened rather than weakened.[4] Evolutionary theory, it was argued, filled in some of the gaps in our knowledge of how God pulled off the mammoth task of creating the

---

[1] Cited in F. Darwin (1887b), p. 23.
[2] Ruse (2001).
[3] Cited in White (1896), p. 103.
[4] Brooke (2003).

world. When Darwin died in 1882, he was given a state funeral and buried in Westminster Abbey, alongside Isaac Newton. It is tempting to think that this symbolizes the compatibility of his theory with religious belief.

Of course, we shouldn't underestimate the discomfort that Darwin provoked among the religious. There were plenty of heated reactions to his theory, and some people really were appalled. Many rejected it outright, and many still do. But just as many see no conflict between their belief in God and their belief in evolution. Admittedly, this is sometimes because they haven't given the matter much thought. However, there are plenty of people who have thought a great deal about the subject, and who are certain that evolution can slot easily into a theistic worldview. In this chapter, we'll turn a critical eye to this position.

Religious Darwinians

As a general rule, Darwin kept his views on religion private. He was acutely uncomfortable at the prospect of offending people, especially his wife Emma, who was a devout Christian. Thus it's not clear exactly what his thoughts on the subject were. It appears, though, that when he first formulated his theory of evolution, Darwin was a theist or deist, and that this may still have been the case twenty years later when he unleashed the theory on an unsuspecting world. Later in life, he generally identified himself as an agnostic, although he seemed to vacillate somewhat and at times would perhaps have been better classed as an atheist. One thing is clear, though: throughout the course of his lifetime, Darwin became less and less religious.

Some critics of evolutionary theory pounce on this breakdown in the man's faith and try to attribute it to the pernicious effects of his theory. But this seems unlikely; the erosion of Darwin's belief in a benevolent deity appears to have been caused as much by family tragedies (in particular, the death of his ten-year-old daughter) as it was by the concept of natural selection. Thus, even though Darwin himself was not a theist for much of his adult life, he certainly didn't think that evolution ruled out theistic belief. In the *Origin*, he wrote: 'I see no good reason why the views given in this volume should shock the religious feelings of any one', and elsewhere he wrote that one could be 'an ardent Theist & an evolutionist'.[5] Plenty of subsequent evolutionists have proved Darwin's point. Consider the following quotations:

> To the traditionally religious man, the essential novelty introduced by the theory of the evolution of organic life, is that creation was not all finished a long while ago, but is still in progress, in the midst of its incredible duration. In the language of Genesis we are living in the sixth day, probably rather early in the morning, and the Divine artist has not yet stood back from his work, and declared it to be 'very good'.

> Evolution is God's, or Nature's, method of creation. Creation is not an event that happened in 4004 BC; it is a process that began some 10 billion years ago and is still under way[6] ... Does the evolutionary doctrine clash with religious faith? It does not. It is a blunder to mistake the Holy Scriptures for elementary textbooks of astronomy, geology, biology, and anthropology. Only if symbols are construed to mean what they are not intended to mean can there arise imaginary, insoluble conflicts.

> God, who is not limited to space and time, created the universe and established natural laws that govern it. Seeking to populate this otherwise sterile universe with living creatures, God chose the elegant mechanism of evolution to create microbes, plants, and animals of all sorts. Most remarkably, God intentionally chose the same mechanism to give rise to special creatures who would have intelligence, a knowledge of right and wrong, free will, and a desire to seek fellowship with Him.

---

[5] The first quote is drawn from Darwin (1859), p. 452. The second is cited in Desmond and Moore (1991), p. 636.

[6] Current estimates suggest that the universe began 13.7 billion years ago; that the earth and the solar system formed around 4.6 billion years ago; and that life on earth began 3.5–4 billion years ago (probably closer to 4 billion).

The first quotation came from Sir Ronald Fisher, the second from Theodosius Dobzhansky, and the third from Francis Collins.[7] Fisher and Dobzhansky were both important evolutionary theorists; indeed, both would make most people's list of the top ten evolutionary theorists since Darwin. As for Collins, he led the team that first mapped the human genome. So these are not just rank-and-file scientists; they're major-league players – and they're also 'ardent theists'. Of course, plenty of distinguished evolutionary biologists have been atheists as well. But this doesn't undermine the point I'm making, which is simply that acceptance of evolution does not *necessarily* imply the rejection of God. Many people have found ways to reconcile evolutionary theory with theistic belief.

At first glance, it's not that hard to do. The core of Darwin's theory (i.e., the idea that species change over time and that all life on this planet is genealogically related) is only incompatible with a literal interpretation of the creation story in Genesis. Anyone willing to accept that the biblical stories are metaphorical could maintain that God created life *through* the process of evolution. Now to some this might seem like a rather obvious and rather lame ploy to defend the Bible from the threat posed by evolutionary theory. It is a widespread view that, before Darwin, everyone was a Young Earth Creationist and took Genesis literally, but that after Darwin believers conveniently discovered that Genesis was intended metaphorically all along. This isn't a fair characterization, however. Non-literal interpretations of the Bible have a long and distinguished history.[8] Saint Augustine, for instance, suggested that the early books of the Bible were written in metaphorical form to make them accessible to an uneducated audience.[9] In other words, Genesis is a dumbed-down account of

---

[7] Fisher (1947), p. 1001; Dobzhansky (1973), pp. 127, 129; Collins (2006), pp. 200–1.

[8] Ruse (2001) provides an enlightening discussion of this issue.

[9] Awkwardly, it's mainly the more educated who interpret these texts metaphorically.

the creation, not a literal retelling. Remember, Augustine lived in the fourth century after Jesus. Thus, today's biblical literalists are even further behind the times than we might have guessed! In any case, as long as one adopts a non-literal interpretation of Genesis, there is no immediate inconsistency between theistic belief and evolutionary theory. The door is open to frame a worldview that incorporates both God and evolution.

The simplest, most obvious, and most common approach is to assume that God steered the course of evolution. There are several variations on this theme. Some theists, for example, imagine that God actively intervened at each step on the path. Others, on the other hand, imagine that evolutionary history is the unfolding of a pre-existing plan residing in the eternal mind of God. What these and related suggestions have in common is the notion that the entire course of evolution was in some way overseen or preordained by God.

Another popular approach is to assume that, rather than personally guiding the process, God chose natural selection as his means of creating life by proxy. A common refrain among advocates of this position is that it was much cleverer of God to have created the biosphere and all its inhabitants through inviolable physical laws than it would have been to do so through piecemeal acts of creativity. The fact that God was able to accomplish so much with such a simple law as natural selection is testament to his genius. As Darwin reported in the *Origin*:

> A celebrated author and divine [Charles Kingsley] has written to me that he has gradually learned to see that it is just as noble a conception of the Deity to believe that he created a few original forms capable of self-development into other and needful forms, as to believe that He required a fresh act of creation to supply the voids by the action of His laws.[10]

So we have two main approaches to reconciling God and evolution. These approaches can be viewed as extremes on a continuum. At one

[10] Darwin (1859), p. 452.

end is the view that God guided the entire evolutionary process and that natural selection played no role whatsoever. At the other is the view that God left everything up to natural selection and did not intervene in any way, shape, or form. Many believers fall somewhere between these two extremes, sharing the workload between God and natural selection. Some allow only a very small role for natural selection. Thus, some ID theorists, for example, grudgingly concede that natural selection can produce trivial changes within species, but insist that God is responsible for everything else (i.e., complex adaptations and the transformation of one species into another). Others give natural selection a starring role; they suggest that, for the most part, God leaves things up to natural selection, but that his guiding hand is needed once in a while to keep things moving in the right direction or to accomplish things that selection alone could not. They suggest, for example, that we must invoke God to account for the origin of life from non-life, or the first DNA, or the first cell, or the first multicellular organisms, or the 'abrupt' explosion of animal life during the Cambrian period, or the first members of each major taxonomic class (e.g., mammals, birds), or the first human beings, or the first human minds. Individuals in this camp are willing to credit the bulk of evolutionary change to natural selection – but not all of it. They insist on leaving room at key junctures for divine fiddling.

There is one other attempted reconciliation worth mentioning. It differs from those discussed already in that it doesn't involve a definite distinction between things done by natural selection and things done by God. Instead, God and natural selection are intimately involved at every step on the path. The idea, which was particularly popular in the immediate aftermath of Darwin's theory, is that God's role in evolutionary history was to supply the variation with which natural selection works. This proposal is most closely linked with the nineteenth-century Harvard biologist, Asa Gray, who was Darwin's main early defender in the United States. Gray denied Darwin's claim that new variants are a product of chance alone, noting that 'we should advise Mr Darwin to assume, in the philosophy of his hypothesis, that variation has been led along certain beneficial lines'.[11] Recently, the ID theorist Michael Behe has attempted to resuscitate this position.[12]

Whatever the details happen to be, the idea that God played some kind of role in evolutionary history – a position known as *theistic evolution* – is very popular. It is common among Christians and Jews, and a number of Muslim scholars have embraced it as well. Unlike creationism, theistic evolution has the support of many scientists and other notable thinkers,[13] and in that sense can be considered an intellectually respectable position. As such, we can't just reject it out of hand. In the remainder of the chapter, we will make a careful examination of the arguments, and *then* reject it.

Not so fast

At first glance, the idea that God masterminded the evolutionary process seems like a straightforward and unproblematic way to reconcile evolutionary theory and religious belief. When we look more closely, though, it turns out that every attempt to effect this reconciliation does grievous damage either to the religion, or to evolutionary theory, or to both. We'll start with religion. The first point to make is that Darwin's ideas have forced a radical rethink of Christian doctrines and theistic belief in general, pushing believers to come up with ever more sophisticated rationalizations for their beliefs. After Darwin, it suddenly takes a lot more intellectual gymnastics to square science with theistic belief. People sometimes claim that science and religion deal with different and distinct subject matter; they are, in

[11] Gray (1888), p. 148.   [12] Behe (2007).

[13] These include Ayala (2007); Collins (2006); Conway Morris (2003); Dobzhansky (1973); Dowd (2007); Fisher (1947); Haught (2000); McGrath (2005); Miller (1999); Peacocke (2004); Polkinghorne (1994); Roughgarden (2006); Swinburne (2004); Teilhard de Chardin (1959); Ward (1996).

Stephen Jay Gould's famous words, *non-overlapping magisteria*.[14] But in as much as this is true at all, part of the reason it's true is that evolutionary theory has forced religion out of its arena. It has led many people to drop aspects of their religion that clash too obviously or too violently with the Darwinian worldview. And it has necessitated a non-literal interpretation of the biblical stories.[15] This is not to deny that some people interpreted the Bible metaphorically before Darwin. However, evolutionary theory made this type of interpretation more urgent and more widespread, and it strengthened people's doubts about the authority of the Bible.[16] Today, every scientifically literate priest or vicar will tell you that the Bible is not a science textbook and shouldn't be treated as one; that it is not intended to tell us *how* the universe was created but only 'whodunit'; or that its true purpose is to tell us where we're going rather than where we came from. But how many people said things like this *before* Darwin provided an alternative explanation for how species came to exist? One might argue that evolutionary science has helped to show us how we should understand the true message of the Bible. Isn't it just as plausible, though, that the Bible's authors were simply wrong?

We do have to concede one thing: the less literally we take the Bible, the more we can reconcile its claims with evolutionary theory. The problem is that the less literally we take the Bible, the more we can reconcile its claims with absolutely anything at all – a fact that hardly enhances the Bible's credibility. Furthermore, not only the

---

[14] Gould (1999).

[15] At any rate, it has necessitated a non-literal interpretation of those stories that clash with evolutionary theory. This is not generally extended to the entire Bible; many Christians still believe, for example, that the story of Christ is literally true. By the way, next time someone tells you that the biblical stories are metaphorical, pick one and ask them to tell you what the metaphorical meaning is. Nine times out of ten they won't have an answer.

[16] Similarly, the view that faith rather than reason is the path to true knowledge of God was held long before Darwin promulgated his theory. But the theory provided all the more reason to jettison reason and evidence, and to base one's beliefs entirely on faith (i.e., on nothing).

biblical creation story but *any* creation story can, if taken metaphori-cally, be reconciled with evolution. And as soon as we take off our cultural blinkers and start to contemplate the creation stories of other cultures, we face some tough questions. Are *all* the creation stories poetic descriptions of the Big Bang and the evolution of life? Are they all moral allegories, rather than stories that people genuinely believed were true? Or are we just making an exception for the creation story we happen to have been brought up with?

The idea that the biblical stories are symbolic is charitable to the point of absurdity. What would we think of a university professor who, happening upon unambiguous errors in a favourite student's work, concluded that the student was speaking symbolically and awarded her top marks? The whole notion that Genesis is metaphor-ical, and that evolution is a testament to the glory of God, smacks of the kind of spin doctoring that gives politicians a bad name. Liberal Christians alter their original religious beliefs to make them compat-ible with evolutionary theory, and then scoff at the idea that there was ever any threat. In doing so, they casually downplay just how radically they've rewritten their religion. Arguably, it is not the same religion as the one it evolved from; it merely shares the same name.

In some cases, people have rewritten their belief systems so extensively that it's a stretch to think that the beliefs they profess can even be considered *religious* beliefs anymore. Earlier, we heard that some of the most distinguished evolutionary biologists in the history of the field have been devoutly religious people. If we look a little closer, though, things are not quite so clear-cut. In a eulogy for Theodosius Dobzhansky, his former student Francisco Ayala noted that 'Dobzhansky was a religious man, although he apparently rejected fundamental beliefs of traditional religion, such as the existence of a personal God and of life beyond physical death.'[17] To my mind, this sounds like a strange way of saying that

---

[17] Ayala (1977), p. 9.

Dobzhansky was *not* a religious man. More importantly, it exemplifies one of the common outcomes of trying to mesh one's religious beliefs with the scientific worldview: you end up doing major violence to the religious beliefs.

The damage goes both ways. Many ostensible efforts to square Darwinian evolution with theistic belief actually involve the rejection of Darwinian evolution. The idea that God personally guides the evolutionary process, for example, is consistent with the bare *fact* of evolution – the simple, unadorned fact that species change over time and that all life is related. However, although it's consistent with the fact of evolution, it conflicts with a central tenet of the modern *theory* of evolution, namely, that the design found in organisms is a product of the mindless (but non-random) accumulation of random (but useful) accidents. Imagine someone who accepted the basic fact of gravitation (e.g., the fact that things fall when dropped), but who didn't accept Einstein's theory of gravitation (i.e., general relativity). We would never say that this person had accepted modern physics into his view of the world. But nor have those who accept the fact but not the theory of evolution accepted modern biology.

Nonetheless, it's a common position. Recall Asa Gray's idea that God provides the variation with which natural selection works. This involves acceptance of the *fact* of evolution, but not of Darwin's *theory* about its causes. Although Darwin didn't know about genes or mutations, and therefore didn't know how new variants came about, he correctly surmised that the process was unguided and random. Indeed, this assumption is integral to his theory; as he himself pointed out, if new variants were sufficiently suitable, there would be nothing left for natural selection to do. As far as Gray was concerned, the denial of intentional design in nature was tantamount to atheism. However, the denial of intentional design is the *defining feature* of Darwin's theory. Thus, Gray's perspective does not reconcile God with Darwinian evolution; it represents the rejection of Darwinian evolution.

What do the major Christian denominations have to say on the subject? Most appear to be at ease with evolution. It is widely believed, for example, that the Catholic church has fully embraced and accepted Darwin's theory. In 1996, Pope John Paul II acknowledged in a papal encyclical that there was no question anymore that evolution had taken place.

> New knowledge has led to the recognition of the theory of evolution as more than a hypothesis. It is indeed remarkable that this theory has been progressively accepted by researchers, following a series of discoveries in various fields of knowledge. The convergence, neither sought nor fabricated, of the results of work that was conducted independently is in itself a significant argument in favor of this theory.

This is a widely quoted passage, and those quoting it – evolutionists and liberal Catholics alike – take great pride in the fact that the pope accepted evolutionary theory. But did he really? Well certainly he accepted the *fact* of evolution. Again, though, he did not accept the modern theory – not completely anyway. In the same encyclical, Pope John Paul denied that natural selection could explain the evolution of humans from non-human apes, an event which supposedly involved a 'transition to the spiritual'. He also denied that natural selection could account for the origin of the human soul (by which I presume he meant the conscious mind, the part of the person that allegedly survives death). So, although it is widely believed that Pope John Paul accepted evolutionary theory, he did not, anymore than someone who thinks that all the planets *except the earth* orbit the sun has accepted the Copernican vision of the solar system.

Continuing this tradition, John Paul's successor, Pope Benedict XVI, asserted that naturalistic approaches to the origin of life and species are incomplete, and that a role must be reserved for God. Other Catholics take a harder line. In 2005, Cardinal Christoph Schönborn, archbishop of Vienna, stated that 'Evolution in the sense of common ancestry might be true, but evolution in the neo-Darwinian sense – an unguided, unplanned process of random variation and natural selection – is not.' In other words, Schönborn openly rejects the modern theory of evolution in its entirety. Contrary to what you might have heard, then, there is no general consensus in the church about the validity of evolutionary theory. Some accept it without reservation, but many do not. The same is true of other major Christian denominations. During the controversy stirred up by Schönborn's statement on evolution, the Protestant theologian Alvin Plantinga – perhaps the most famous and influential theologian living today – expressed his agreement with Schönborn's position. In other words, Plantinga accepted the possibility that life had evolved but rejected the modern theory of the causes of evolution. What this and the earlier examples show is that many efforts to reconcile God and evolution involve rejecting evolutionary theory. It is not as easy to reconcile God and Darwin as we might previously have thought.

What's wrong with theistic evolution?

We can concede this much: theistic evolution does provide a solution to the God v. Darwinian evolution dilemma. But the solution is not to reconcile these views; it is to reject Darwinian evolution. Theistic evolution and Darwinian evolution are, quite simply, inconsistent with one another. Of course, this in itself does not automatically imply that theistic evolution is false. So our next question is: are there any reasons to opt for Darwinian evolution over theistic evolution? It turns out that there are, and that they're very good ones.

Evolutionary biology raises major problems for the view that God personally guided the process of evolution, or even that he did so to any great extent. One problem stems from the fact that adaptations appear to be designed to enhance inclusive fitness – in other words, to increase the likelihood that the genes contributing to their development will be passed on, either by the individual possessing the

adaptation or by that individual's genetic relatives.[18] It is curious that
God has used inclusive fitness as his guiding principle, for this means
that he chose to design life by a criterion that would make it perfectly
explicable as the product of a mindless, unguided process of gene
selection. Why would he do this? Why would he hide any evidence of
the hand he had in directing the great drama of life?

A second and more serious problem is that in every member of
every species on this planet we find design flaws and imperfections,
and, as David Hume argued centuries ago, imperfections in design
constitute dubious evidence of a perfect designer.[19] The best-known
example is the fact that in humans and other vertebrates the eye is
wired up backwards, making our vision less detailed than it would
otherwise be and giving us a blind spot in our field of vision. I'm not
suggesting for a minute that the eye is *badly* designed; it's extremely
well designed and it serves us very well. But there are two points to
make. First, if it had been wired up the right way round, it would have
been slightly better at no extra cost. Second – and this is really the
clincher – if the eye had been wired up the right way round, it would
have been so much more elegant a design, so much more like something an intelligent being would have put together. Instead, it looks
exactly as we'd expect if it were the product of a mindless, unguided
process. There are many other examples of what we might call *unintelligent design*. Many of us have too many teeth in our heads given the
size of our jaws, which is why we often have problems with our wisdom
teeth. Ironically, then, our wisdom teeth illustrate not the wisdom of
our 'creator', but his incompetence. We are vulnerable to haemor-
rhoids, varicose veins, and lower back pain, all of which is a result of
the fact that we are not perfectly adapted to walking upright – not
surprising when you consider that we evolved to walk on two legs

[18] On inclusive fitness theory/the genes'-eye view of evolution, see, e.g., Cronin
(1991); Dawkins (1982, 1989); Hamilton (1963, 1964); Williams (1966). See
also the discussion of kin selection in Chapter 11.
[19] See Gould (1980); Miller (1999); Shubin (2008).

only very recently in evolutionary time, and that for millions of years before that our ancestors walked on all fours. There are many other design flaws and imperfections in the biological world, and they all raise the same question: why would a perfect creator design such imperfect beings as ourselves?

This is a hard question to answer. In contrast, it's very easy to understand why natural selection can never produce perfection. First, it is a process without foresight, limited to 'choosing' the best (or at any rate, the least bad) options from among the variants thrown up by pure chance. And, second, every step in the gradual evolution of any adaptation has to be advantageous to the genes involved. Many sensible designs cannot be reached because there is no path to them through an unbroken series of advantageous steps. Imperfections in the design of living things are therefore exactly what we'd expect if life is a product of natural selection, rather than the creation of God, whether through instantaneous creation or guided evolution. Admittedly, intelligent design does not imply *optimal* design. But the problem isn't just that the designs aren't perfect. The problem is that they're imperfect in ways that make sense only on the assumption that they are products of natural selection, unaided by any intelligent force.

One might object that it is possible in principle that the imperfections we perceive are only apparent, and that they're actually there for some purpose we cannot fathom. The ID theorist Michael Behe observed that: 'Features that strike us as odd in a design might have been placed there by the designer ... for artistic reasons, for variety, to show off, for some as-yet undetected practical purpose, or for some unguessable reason.'[20] God moves in mysterious ways, in other words. The problem with this kind of argument – known in the philosophical literature as a *plea from ignorance* – is that it can be used to sweep under the carpet any possible inconsistency or awkward fact. As such, it's hardly a compelling move. When you have to resort to this kind of

---

[20] Behe (1996), p. 223.

argument, it shows, at the very least, that your theory does not do a very good job of explaining your data; if it did, you would have no need to employ the argument. But it suggests more than this. It suggests that you're grasping at straws. When the best you can do is resort to a plea from ignorance, perhaps what you *should* be doing is conceding that you've lost the argument.

Is there any way that the Theistic Evolutionist can evade these criticisms? There is. The arguments considered above apply only if God's guiding hand is the sole force, or at least the main force, underlying evolution. If we allot God a vastly reduced role in directing evolutionary history, and give most of the credit to natural selection, we have a ready explanation for the fact that adaptations are designed to enhance inclusive fitness, and for the fact that the biological world is jam-packed with imperfections. At first glance, this seems to be an attractive solution. It basically accepts Darwinian evolution but still retains a small role for God. However, it turns out that the so-called solution simply exchanges one set of problems for another. First, if God is *ever* willing and able to intervene, why doesn't he intervene to iron out some of the imperfections in the design of his favoured creations? In particular, why doesn't he get rid of those that cause needless suffering? If he is capable of intervening to produce things like bacterial flagellar motors, why can he not also intervene to produce immunity to malaria in African populations or to neutralize the dangers posed by the appendix? Is God more interested in mass-producing nifty bio-machines than in reducing the suffering of human beings and other animals? (I'll say more about the problem of suffering in Chapter 6; it is a major stumbling block for this brand of theistic evolution.) Furthermore, as long as God is held to intervene at all, we're left with an awkward implication, namely, that although natural selection is good enough to accomplish many things, God still has to step in and help out with some especially tricky designs. Doesn't this suggest that the natural laws he established for the universe are imperfect for bringing about the things he desires? Doesn't it suggest that his chosen method of creation isn't quite good enough to do everything he wants it to do, and that he must therefore break his own rules to keep things working according to plan? Doesn't it suggest a less-than-perfect deity?

Deistic evolutionism

The obvious solution to this little conundrum is to reduce God's role in guiding evolution to zero – to assume that God chose to create life entirely via the mechanism of natural selection. We'll call this position *deistic evolution* to distinguish it from theistic evolution, which involves at least some direct intervention from God. Deistic Evolutionists hold that God created the universe and the laws of nature (and perhaps also that he jump-started life), but that once the ball was rolling, he ceased to intervene in the day-to-day running of the world or in the course of natural law. God was like the ether after Einstein: he no longer had any role to play in the universe. Deistic evolution sidesteps the problems we've considered so far. First, it is perfectly consistent with Darwinian evolution; second, it can accommodate design flaws and the fact that adaptations are designed to maximize inclusive fitness; and third, it does not imply that God's method of creation was inadequate for designing life. Beyond this, deistic evolution eliminates any immediate conflict between science and belief in God. Anyone who believes that God's role was merely to create the laws of nature can accept the scientific worldview in its totality; they simply add the proviso that 'God did it' – i.e., that God is responsible for the world that science describes. On all these points, deistic evolution trumps theistic evolution.

It's an impressive résumé, and not surprisingly the position has attracted a number of respected supporters. At times, Darwin flirted with deism. He was convinced that evolution was an entirely mindless and unguided process, but he was sympathetic to the possibility that this mindless process might itself have been designed. Occasionally, he entertained the idea that there was a God, but that God's role in the creation of life was simply to set up the arena in which natural selection took place (i.e., the physical universe). In a letter to the American biologist Asa Gray, shortly after the publication of the *Origin*, Darwin wrote:

> I cannot anyhow be contented to view this wonderful universe, and especially the nature of man, and to conclude that everything is the result of brute force. I am inclined to look at everything as resulting from designed laws, with the details, whether good or bad, left to the working out of what we may call chance. Not that this notion at all satisfies me. I feel most deeply that the whole subject is too profound for the human intellect. A dog might as well speculate on the mind of Newton.[21]

More recently, some major modern evolutionists, including E. O. Wilson, have accepted the possibility of deism. But although the attraction is clear, there is a high price to be paid for adopting this position. Deistic evolution strips God of many attributes that most believers consider central. It rules out a God that gifts us with revelations or authors books. It rules out a God that defies his own laws and performs miracles. And it rules out a God that intervenes in people's lives or answers prayers (unless you can believe that God leaves the evolution of all organic life entirely unattended, but then ministers to the fleeting wishes of individual members of the species *Homo sapiens*). These are implications that most religious people will be unwilling to accept.

And so they should be. The God of deism is emphatically not the God of the monotheistic religions or the God that most people believe in, and we must not gloss over this fact. If you are a deist, you cannot claim to be a believer in Judaism, Christianity, or Islam. In fact, it's not clear what you really *do* believe. The deist's God has much less content than the more traditional conceptions. Thus, those who make a sincere effort to hang onto both God and Darwin end up

[21] Cited in F. Darwin (1887b), p. 312.

with a highly circumscribed and anaemic version of the religion to which they once belonged. One might even argue that they are actually in the process of losing their religion. They not only drain most of the colour out of their God but they render God superfluous. As such, they should ask themselves: if the universe looks exactly the same with or without God, what reason is there to believe in God? Once again we see that, regardless of the details of the reconciliation, it is extremely difficult to incorporate an incorporeal creator into a worldview informed by Darwin.

Conclusion

From the moment Darwin's dangerous idea caught hold of the public imagination, believers have been trying to reconcile their faith with the reality of evolution. If we permit a non-literal interpretation of the Bible, and if we squint hard enough, it is possible to reconcile Genesis with evolutionary theory, and after that it's easy to weave a narrative containing both God and evolution. In this chapter, we looked at two such narratives: the idea that God personally guided the evolution of life, and the idea that he invented natural selection as his means of creating life by proxy. But although both ideas look inviting on the surface, as soon as we dig a little deeper, we find that the attempted reconciliations do an injury either to evolution, or to theism, or to both. Evolutionary theory renders a hands-on God much less plausible than he might once have been. It puts pressure on thoughtful believers to abandon theism, and pushes them towards either deism or atheism. How can we choose among these remaining options? In the next chapter, we'll evaluate – and ultimately reject – some of the best arguments put forward for adopting deism, and we'll consider what evolutionary theory contributes to the issue.

FIVE

God as gap filler

Darwinism can only explain why some animals are eliminated in the struggle for survival, not why there are animals and men at all with mental lives of sensation and belief; and in so far as it can explain anything, the question inevitably arises why the laws of evolution are as they are. All this theism can explain.

Richard Swinburne (2002)

Anything you don't understand, Mr Rankin, you attribute to God. God for you is where you sweep away all the mysteries of the world, all the challenges to our intelligence. You simply turn your mind off and say God did it.

Dr Arroway in Sagan (1985), p. 166

Not how the world is, is the mystical, but *that* it is.

Ludwig Wittgenstein (1921), 6.44

Life, the universe, and everything

We've seen that there is no role for God within the evolutionary process. But perhaps a role can be found for him outside that process. Darwin might have explained the origin of species, but there are still many mysteries left that believers claim support the existence of God. In this chapter, we'll consider three. First, natural selection can only occur when there is something to select, and Darwin's theory says nothing about how life began in the first place. So there we have a possible role for God. Second, even if evolutionary theory and science in general explain much of what we find within the universe, we have yet to explain the fact that the universe exists at all. Not only that, but the universe appears to be perfectly tailored for life: if the laws of nature were even slightly different, life (or at least life as we know it) could never have evolved. Perhaps, then, we must posit God as initiator and fine-tuner of the universe. And third, although evolutionary theory explains the existence of human bodies, some argue that human consciousness is not amenable to a naturalistic explanation and must therefore be attributed to God. In each of these areas, the theist could argue that we have a role for God that Darwin cannot usurp.

Much of the discussion of these issues is unconnected with evolutionary theory, and is thus beyond the scope of this book. However, the theory does help to challenge the need for religious explanations in each of the new territories that theists have staked out, and in the process it weakens each of the remaining arguments for the existence of God. Some claim that life, the universe, and mind are miraculous. The conclusion of this chapter will be that these things are really, really amazing, but that they're not miraculous (unless by miraculous, you just mean 'really, really amazing'). We'll start with the origin of life.

Life from non-life[1]

One of the great mysteries of the universe is captured in the following question: which came first, the chicken or the egg? If chickens only come from eggs, and eggs only come from chickens, how did the whole process begin? Most people first encountered this question in early childhood, and gave up trying to answer it soon afterwards. To

---

[1] Parts of this section are based on Stewart-Williams (2003).

all those who have wrestled with the question, I am pleased to announce that a definite answer can now be given: *the egg came first*. To see why, we need only observe that chickens evolved from a pre-existing, egg-laying bird species.

But of course I'm deliberately missing the point. The real question is: how did life begin? If new life comes only from existing life, how did the whole cycle get started?[2] Darwin's theory beautifully explains change within species and the origin of new species. However, natural selection can only occur once self-replicating life already exists. Evolutionary theory is silent on the issue of how life first came about. The working hypothesis made by scientists in the area is that the origin of life, like the origin of species, was a strictly natural and unguided process. Some religiously motivated critics emphatically deny this possibility. Life, they claim, could not possibly have come about through the natural workings of the laws of physics and chemistry; to suggest that it did is simply not credible. Those making this assertion are often Creationists, and they tend to view it as a powerful argument against evolutionary theory. It's important to remember, though, that the question of the origin of life is distinct from the question of the evolution of species. Even if the Creationists are right that life could not emerge spontaneously from non-life, we already know for a fact that the different species evolved. The only question now is what happened before biological evolution began. Did God create the first self-replicating entities or did they come about through natural processes?

Although it's hard to imagine how life could form from non-life without intelligent intervention, various scenarios have been proposed. An early effort to formulate a naturalistic account of life's origins was the theory of *spontaneous generation*. According to this ancient

---

[2] There are several very good introductions to the topic of *abiogenesis*. These include Cairns-Smith (1985); Davies (1999); Hazen (2005); Maynard Smith and Szathmáry (1999).

theory, life arises spontaneously and regularly from non-living matter. To early observers, there seemed to be good evidence for this. It appears, for instance, that maggots form spontaneously from rotting meat, and that mice form from stored grains. We now know that maggots and mice come only from the fertilized eggs of adult members of their respective species. Thus, we now know that spontaneous generation does not happen, at least not in the way it was once thought to. However, anyone who accepts that the emergence of life can be understood in purely naturalistic terms is committed to the view that the spontaneous generation of life from non-life happened at least once. It would not have involved mud suddenly morphing into mice or men, though; it would have been an extremely slow process and the first life forms would have been far simpler than any we know today. How did it happen?

Darwin said nothing publicly about how life might have begun, apart from a passing statement in the *Origin* (perhaps intended to placate religious readers), that life was 'originally breathed by the Creator into a few forms or into one'.[3] In a private letter, though, he speculated that life had begun in 'some warm little pond, with all sorts of ammonia and phosphoric salts, light, heat, electricity, &c., present, [so] that a proteine compound was chemically formed ready to undergo still more complex changes'.[4] Subsequent theorists started adding meat to this skeletal suggestion, outlining the basic sequence of events necessary for life to arise spontaneously from non-living matter.[5] There are three steps. First, you need a supply of organic molecules, such as amino acids and nucleic acids. These are the bricks and mortar of the organic world, the basic building blocks of life. Second, these organic molecules must coalesce into long, complex chains, or *polymers*. Third, and finally, these polymers must be

---

[3] Darwin (1859), p. 459.
[4] Cited in F. Darwin (1887b), p. 18.
[5] Haldane (1929); Oparin (1924); Schrödinger (1944).

organized into a self-replicating structure – an entity capable of making high-fidelity copies of itself. As soon as replication arrives on the scene, Darwinian evolution takes over and the rest is history.

Needless to say, I'm making a very long story very short, so let's consider each step in turn and ask whether any of them demand intervention from on high. The first step is unproblematic: experiments have shown that organic molecules can form from inorganic molecules through entirely natural processes.[6] These molecules could have formed on earth, or the earth could have been impregnated by a rain of organic materials formed in space. Either way, there's no need to invoke God for the first step. The same applies to the second; chains of organic molecules could have formed on clays or through other means.[7] But getting this far is the easy part. The hard part is explaining how, over millions of years, these chains came to be organized in such a way as to form the first self-replicating structures. How did we get from a world littered with organic molecules to a world in which Darwinian natural selection could take over? This is where unguided naturalistic processes begin to seem inadequate. The chances of a self-replicating molecule forming completely by accident seem impossibly slim – as slim, according to the standard metaphor, as the chances of a tornado whipping through a junkyard and assembling a jet airplane. So how could it happen?

There are three main responses to this challenge. Response number 1 is that the spontaneous formation of a self-replicating molecule might not be as unlikely as our intuitions tell us it is. The fact that it seems so utterly implausible may stem in part from our tendency to think in categorical terms. We naturally divide the world into living

---

[6] The first research in this area was done by graduate student Stanley Miller (1953), working with the chemist Harold Urey. Their early experiments were flawed in a number of ways. Nonetheless, further investigation has revealed that, given the right circumstances, many of the molecules essential to life form naturally and inevitably.

[7] See, e.g., Cairns-Smith (1985).

things and non-living things, and most of the time this dichotomy serves us well. It may, however, tempt us into imagining that there is an unbridgeable gulf separating inanimate matter from the first life forms. Darwinian principles suggest an alternative way of thinking about this. In a very real sense, there *was* no first life form. At one time in the earth's history, there was clearly and unambiguously no life. At a later time, there clearly and unambiguously was. In between, however, there were intermediate forms that could not be classed as living or non-living – not without choosing a wholly arbitrary cut-off point. This idea that there were intermediate stages between life and non-life is supported by the fact that there are entities in the modern world that fall into this conceptual no-man's land: viruses. As soon as we start thinking in terms of a continuum between non-life and life, with no cut-off point between these two states of matter, we lessen the perceived problem of how life could form spontaneously from non-living materials. It seems as though there must be a giant leap at some point, but perhaps that's not the case.

A second point is that, even if the spontaneous formation of self-replicating molecules really *was* a cosmic fluke, we would still have no reason to think that it was anything other than a naturalistic process. The universe is very big and very ancient. As such, even the most improbable events are likely to happen at least once in a while. As the biologist Richard Dawkins has pointed out, even if the odds of life forming spontaneously are a mere one in a billion, there are at least a billion, billion planets in the observable universe alone, and we would therefore expect life to have come about roughly *a billion times*, purely by chance.[8] Here's another way to think about it. The likelihood that you will meet the love of your life today is very low. But the likelihood that you will meet the love of your life on *one* of the many days of your life is very high (most people do). In exactly the

[8] Dawkins (2006).

same way, the likelihood of life forming spontaneously on any given planet is very low – so low as to seem miraculous. However, if we take into account the fact that there are billions upon billions of planets, it would be miraculous if life had *not* evolved somewhere in the universe. It may be a near certainty that life will *not* evolve on any *particular* planet, but a near certainty that it will evolve on *some* planet. This might sound like a contradiction; however, it is a simple statistical consequence of the vast number of planets in this universe. What this tells us is that, even if we concede that the spontaneous generation of life really is ridiculously improbable, we still have no reason to invoke God.

The third point is one we touched on in Chapter 2, but it's a point worth repeating. No one should deny that the origin of life is a profound mystery, one that screams out for an answer. The order and complexity found in living things is amazing and incomprehensible. However, to say that life was created by God does not solve the mystery; it magnifies it. It accounts for the existence of life, but it does so by positing something even more mysterious, more amazing, and more difficult to explain: a being capable of creating life. Theists point out, quite reasonably, that the origin of life through natural processes seems incredibly unlikely. But their attempted solution is to suggest something even *more* unlikely. Positing God does not solve the mystery of life's origins, and therefore the mystery of life's origins provides no reason to posit God. This doesn't disprove the existence of God, but it does undermine this particular argument for God. It therefore leaves the God hypothesis on thin ice, and raises again that menacing question: what reason do we really have to believe?

Although the exact details of the evolution of life from non-life may always remain a mystery, research in this area shows at least that there is no need to think it was anything but a natural process. It is only reasonable to assume that life developed from non-living matter through a completely natural process, a process in which chemistry slowly evolved into biology. After a long period of chemical evolution, biological evolution began, eventually producing all the billions and billions of species that have inhabited this planet – including one strange species that can ponder issues such as these.

The origin of the universe

The argument that God must have kick-started life fails. But perhaps we can push God's role back further in time. Even if scientists can explain how life came to exist in the universe, they have yet to explain how the universe itself came to exist. Maybe we need to posit God to answer this most profound of questions. Here's a simple version of the argument, one more likely to be found on the lips of everyday believers than of professional philosophers: 'Look around you at all the beauty and majesty of this universe. How can you possibly believe that all the amazing order and grandeur could just have sprung out of nowhere? There must be a God that created it all.' A more sophisticated version of the argument, known as the Cosmological or First Cause Argument, runs as follows:

> Every event has a cause.
> The chain of causes cannot stretch infinitely back into the past.
> *Therefore*, there must be a First Cause, and this is what we call God.

As with the origin of life from non-life, Darwin didn't speculate about the origin of the universe in his public writings. Nonetheless, the Cosmological Argument did have some sway over his opinions. In a letter dated 1873, he wrote:

> I may say that the impossibility of conceiving that this grand and wondrous universe, with our conscious selves, arose through chance, seems to me the chief argument for the existence of God; but whether this is an argument of real value, I have never been able to decide.[9]

[9] Cited in F. Darwin (1887a), p. 324.

More recently, the evolutionary theorist, E. O. Wilson, has adopted a similar position. According to Wilson, the existence of a Biological God (i.e., a God that had a hand in the creation of the biological world) has been disproved by evolutionary science. However, the existence of a Cosmological God (i.e., a God that created the universe) is still an open question. In Wilson's view, the question is a problem in astrophysics.[10]

Despite Darwin's and Wilson's guarded endorsement of the Cosmological Argument, the argument is deeply flawed. First and foremost, positing God might solve the problem of how the universe came to exist, but it leaves us with another problem: how did God come to exist? And this is a vastly more difficult problem, because God is vastly more amazing than the universe. If we have to posit a God to explain the universe, wouldn't we then have to posit a Super-God to explain God, a Mega-Super-God to explain Super-God, and so on ad infinitum? The theist will resist this suggestion, and insist that God doesn't require a cause. However, if we can accept that God exists without a cause, why can we not just accept that the universe exists without a cause and leave it at that? It's a lot less to swallow, given that God is infinitely greater than the universe. If it doesn't seem to you that God is the type of thing that requires an explanation, this is because you are tacitly defining God as a being who is somehow the cause of his own existence. But simply defining words in certain ways does not and cannot answer our deepest questions about reality.[11]

Even if the universe as a whole *does* require a cause, it is highly debatable that we should give this the name *God*. As Darwin pointed out, the idea that the First Cause can be identified with God is pure

[10] Wilson (1998).

[11] One approach for the theist might be to argue that it is more plausible that something perfect (God) could exist without explanation than it is that something imperfect (the universe) could exist without explanation. But I see no reason to accept this.

speculation, and it would be more honest to admit our ignorance than to pretend we know something we don't. Of course, identifying the First Cause with God is not pure speculation if one *defines* God as the First Cause. But if that's your definition of God, you cannot then surreptitiously import any other attributes into your God concept. You would have precisely zero reason to think that your God is a personal, creative, conscious being; that it loves you or your loved ones; or that it bears any kind of resemblance to the human mind. You would know virtually nothing about God. So why give it that name? If we called it something else – Alpha, for instance – this would immediately strip away the religious baggage from the concept. However, the new name would also make it much clearer that the notion of First Cause is in fact a hypothesis in physics. And it's a hypothesis that most physicists consider superfluous. Furthermore, whether we call it Alpha or God, the argument still faces various challenges. Just for starters, many of Hume's criticisms of the design argument apply with equal force to the First Cause Argument. How do we know that the chain of causes will trace back to just one cause, rather than two or many? How do we know that the First Cause is not evil? How do we know that the First Cause still exists?

That's a brief survey of the weaknesses of the First Cause argument. Now let's consider what evolutionary theory brings to the party. The main thing to notice is that the entire argument rests on the intuition that the universe must have a cause. An evolutionary perspective on the mind suggests we should be highly sceptical of this intuition. The capacity to construe the world in terms of causes and effects plausibly has an evolutionary origin. This capacity is appropriate and useful within the conditions and circumstances in which it evolved, i.e., the world of medium-sized objects and events. However, beyond these conditions and outside these circumstances, it may simply be inapplicable. This appears to be the case, for example, when it comes to the behaviour of matter at a subatomic level. According to the Copenhagen Interpretation of quantum phenomena (which is probably the most widely accepted view on the topic), some events at the subatomic level *do not have causes*. This includes the timing of the decay of radioactive nuclei and the measured states of electrons and photons. We can assign probabilities to these events, but we cannot predict them precisely – not because we don't understand them well enough, but because they are fundamentally, irreducibly unpredictable. They are literally uncaused events.

Thus, our natural intuition that every event must have a cause appears to be false when it comes to the behaviour of matter at a subatomic level. It may also be false when it comes to the question of the origin of the universe. The fact is that the universe may simply have no cause. This is not just idle speculation. Many physicists believe that the universe emerged from an uncaused quantum fluctuation in the space-time void.[12] A quantum fluctuation is the spontaneous appearance of energy out of nothing. 'Wait a minute', cries common sense (perhaps bolstered by the law of the conservation of energy). 'You can't get something from nothing!' But you can if the something in question *adds up* to nothing. Quantum fluctuations involve the simultaneous appearance of a particle and a matched antiparticle. This happens all the time, and usually these 'virtual' particles collapse straight back into each other and into nothingness. But not always. In at least one case, things worked out a bit differently and, rather than nothingness, we ended up with the Big Bang and the universe. Consistent with this hypothesis, the total energy of the universe appears to be zero. Simplifying somewhat, the physicist Stephen Hawking noted that: 'In the case of a universe that is approximately uniform in space, one can show that [the] negative gravitational energy exactly cancels the positive energy represented by the matter.'[13] In other words, the universe itself is something that

---

[12] See, e.g., Atkins (1993); Craig and Smith (1993); Stenger (2003, 2007); Tryon (1973); Vilenkin (1982, 2006).

[13] Hawking (1988), p. 129.

adds up to nothing. This is consistent with the idea that it was initiated by an uncaused quantum fluctuation – the falling apart of nothing into equal-but-opposite somethings. In everyday life, it is always useful to assume that the things that happen have a cause, and no doubt this was the case for our hunter-gatherer ancestors as well. Perhaps, though, this rule of thumb does not apply to universes. Perhaps things like universes just exist inexplicably. This might seem like a cop-out. But how is it any more problematic than the idea that God just exists inexplicably?

Of course, we still have no idea why there exists a space-time void in which quantum particles and universes can just spring into existence. This brings us face-to-face with one of the most profound questions the human mind can pose: why is there something rather than nothing? But this is not the conversation stopper that the theist supposes it to be. After all, the question applies just as well to God as it does to the universe. Why is there God rather than nothing? And why *this* God rather than another? Invoking God doesn't help solve the mystery of why there's something rather than nothing; it just papers over a gap in our understanding. God becomes a pseudo-explanation for something we simply do not understand. Our overall conclusion? The existence of the universe provides no reason to believe in God.

The Goldilocks enigma

But maybe we can look elsewhere for such a reason. Various aspects of the universe turn out to be surprisingly – one might say suspiciously – suitable for life. If the universe were even slightly different, neither life nor mind could have evolved.[14] To some, this constitutes proof that the laws of nature are what they are *so that* life and mind could evolve, and the obvious inference for people brought up to believe in

[14] See, e.g., Barrow and Tipler (1986); Davies (2006); Leslie (1989); Rees (1999).

God is that God is behind this happy state of affairs. There are two main versions of the argument. The first, which is not very persuasive, is that planet earth is fine-tuned for life. The second, which is more persuasive but ultimately no more successful, cites fine-tuning at a deeper and more fundamental level: the level of the physical laws and fundamental constants of the universe. We'll look very briefly at the first version and then focus on the second.

The first version of the fine-tuning argument claims that, against all reasonable odds, the earth and solar system are set up so that life, mind, and knowledge can exist here.[15] If things were even slightly different, none of these things could have evolved. I'll give you an example. Life as we know it depends on liquid water. If the planet were too hot, all the water would boil away. If the planet were too cold, all the water would freeze. Therefore, a planet hospitable to life must be one that does not get too hot or too cold. This imposes a number of restrictions on the type of planet on which life can evolve. For one thing, the orbit of such a planet must be within a relatively narrow range of distances from its star. This is known as the zone of habitability or – more appealingly – the 'Goldilocks zone': not too hot and not too cold, but just right. The orbit of a life-supporting planet must also be relatively close to circular; if it were too elliptical, the planet would shoot in and out of the Goldilocks zone and life could not be sustained. Another example of apparent fine-tuning is the fact that the earth's gravity is strong enough to prevent the atmosphere from floating away, but not too strong to prevent the emergence of living, moving things. In these and numerous other ways the earth seems to be exquisitely, unexpectedly suitable for life.[16]

How can we explain this peculiar fact? One explanation is that God set things up this way. But there are many objections to this

---

[15] Gonzalez and Richards (2004); Ross (2001).

[16] Richard Dawkins (2006) provides an interesting and more detailed discussion of this issue.

proposal. One is that it has things back to front: the earth is not fine-tuned for life; life is fine-tuned for the earth, and evolutionary theory explains how the fine-tuning took place. A related objection is that although life as we know it could not have evolved if the earth had been much different, life as we *don't* know it quite possibly could have. There's also a statistical point to make. The likelihood that a given planet is habitable may be inconceivably low. As we've seen, though, there are lots and lots of planets! Scientists suspect that there are billions of planets in this galaxy alone. Purely by chance, a tiny minority of these planets will be able to support life. How did we get so lucky as to find ourselves on one of the life-supporting planets, despite their rarity? Easy: if we didn't find ourselves on one, we wouldn't be around to find ourselves anywhere else! This is an application of what is known as the 'anthropic principle': the idea that our theories about the world must be consistent with the fact that we exist in the world. The anthropic principle doesn't *explain* the fact that we find ourselves on a life-supporting planet; what it does is show that no explanation is necessary.

We can safely turn our backs on the fine-tuned earth argument. Let's move on to the second and more powerful variation on this theme: the fine-tuned universe argument. Physicists have identified two dozen or so fundamental physical constants. This includes the speed of light in a vacuum, the gravitational constant, the strong nuclear force, the weak nuclear force, and the proton–electron mass ratio. Some authorities claim that these fundamental constants are poised on a knife-edge, in the sense that if they had even slightly different values, the universe would be utterly unsuitable for the evolution of life.[17] For example, if the force of the Big Bang had been even slightly weaker, the universe would have collapsed in on

[17] Not everyone agrees with this assessment. The physicist Victor Stenger (2003) argues that the constants could be changed quite significantly without rendering the universe unfit for life.

itself long before stars and planets had time to form, and thus long before life could have evolved. On the other hand, if the force of the Big Bang had been slightly greater, the universe would consist of little more than a diffuse cloud of particles, and again there would be no stars or planets – and no life. But instead, the force of the Big Bang happened to be just right to permit the formation of stars and planets, and thus to provide a setting for the evolution of life. Many such cosmic coincidences have been identified,[18] and they all raise the same question: why does the universe appear to be fine-tuned for life?

The (putative) fact that the universe is fine-tuned for life doesn't necessarily imply that God was the fine-tuner, but that's the most common inference. Some take the story a step further, and suggest that God deliberately set up the universe in such a way that intelligent life forms would evolve and eventually start to study the universe, and that in doing so, they would recognize that their existence was due to coincidences in the parameters of the universe's laws and constants. In this way, they would be led by reason to know and love God (the atypically smart ones would, at any rate).

The fine-tuning argument is essentially the argument from design applied to the laws of nature, rather than to the intricacies of biological mechanisms. For my money, it is the most persuasive of the current batch of arguments for the existence of God. But there are good reasons to reject it, and some of these draw on Darwinian principles. An initial point, before we consider anything else, is that even if the fine-tuning argument really does make God more probable than before, this does not mean it makes God's existence more probable than his non-existence. It all depends on how probable God was before we encountered the fine-tuning argument. It seems to me that evolutionary theory renders the idea of God extremely implausible. The fine-tuning argument may reduce this implausibility a little bit, but not to the point where believing in God is actually reasonable.

---

[18] See Barrow and Tipler (1986).

Thus, even if the argument works, it may fail to establish its conclusion. But does it even work? To start with, we need to ask whether the universe really is so suitable for life. Although the universe is bio-friendly, it could certainly be a lot friendlier. More than 99 per cent of the species that have ever existed on this planet are now extinct. As well as a constant trickle of extinctions, there have been at least five mass extinctions during the lifetime of the earth. Up until now, life has always managed to bounce back from the edge. This won't always be the case, though. Life is fighting a constant battle, a battle it must ultimately lose – as the Buddhists say, given infinite time, anything that *can* break *will* break. Taking a wider perspective, most of the universe consists of empty space, and most of the matter of the universe – including most planets – is utterly inhospitable to life. There are quite probably other configurations of the fundamental constants that would allow much *more* life to exist than is found in this universe. Are we to assume that God loves life, but not too much of it?

As for the idea that the universe is fine-tuned for mind or intelligence, Bertrand Russell noted that 'If it is the purpose of the Cosmos to evolve mind, we must regard it as rather incompetent in having produced so little in such a long time.'[19] It might be argued that it is quality rather than quantity that matters, but the cosmos comes up short by that criterion too, at least if local (earthly) trends are anything to go by. And consider this: eyes have evolved at least forty times on this planet alone, and the capacity for flight has evolved numerous times as well. As far as we know, though, humanlike intelligence has evolved only once. Thus, a stronger case could be made that the universe is designed to evolve eyes and wings than that it is designed to evolve intelligence.

There's a deeper problem with the idea that the universe is fine-tuned to produce life. Life is not the only amazing thing that could exist; it's just one that happens to be possible in this universe. For all

[19] Russell (1935), p. 216.

we know, other possible universes would produce things just as amazing as life, if not more amazing – things that we might not even be able to imagine but which are ruled out by the configuration of physical constants in this universe. Our universe is reasonably suitable for life, and life is amazing (in its own opinion, at least). However, the fact that the universe is suitable for something amazing is not necessarily as amazing as it seems, because we are not taking into account all the other potentially amazing things that could exist in other possible universes.

Furthermore, even if the universe really was fine-tuned, how would we know that it was fine-tuned for us? Although it's true that the universe is suitable for the evolution of life, it is equally true that the universe is suitable *for everything else* that exists here. So how do we know that the universe was fine-tuned for life or mind, rather than for aspects of the universe that happen to be of less interest to us? How do we know that it's not fine-tuned for large stars, or the heavy elements of the periodic table, or human technology rather than human minds? How do we know that it's fine-tuned to produce humans in all their glory, rather than just one particular feature of our species? There's a lot of suffering on this planet, made possible by the evolution of a nervous system capable of suffering. For all we know, the universe might be fine-tuned to produce suffering! This might sound silly – and it is. But that's the point. It is no *less* silly to think that the universe is fine-tuned for humans or life or mind than it is to think that it is fine-tuned for suffering. All such claims are equally lacking in evidence. Life and mind seem important to us, but we *are* living beings with minds, so perhaps we're not being impartial about this.

We've seen, then, that even if the universe was fine-tuned, we still have no clear idea what it was fine-tuned for. On top of that, we have no clear idea who or what did the fine-tuning. The evidence is ambiguous with respect to God. The fine-tuning of the cosmological parameters provides just as good evidence for, say, an advanced species of world-creating aliens from another universe as it does for

God. One might ask where the world-creating aliens and their universe came from in the first place. Good question! Here's another good question: where did God come from in the first place? And even if we ignore all these points, we've got to ask whether God actually helps to explain why, out of all possible universes, the one that happens to exist is fine-tuned for humans, life, and mind. Doesn't the God hypothesis simply replace this problem with another: explaining why, out of all possible Gods, the one that happens to exist is the one that decided to fine-tune the universe for humans, life, and mind? The apparent fine-tuning of the universe seems mysterious, and maybe it really is. However, it provides no reason to think that there's a God.

At home in the multiverse

We've considered some strong reasons to reject the fine-tuned universe argument. But we've yet to consider the strongest reason of all. This stems from the opinion of many physicists that this universe is not the only one there is. There is a widespread view now that there may be multiple universes, or even an infinite number of them.[20] This idea, known as *the multiverse hypothesis*, is not as crazy or extravagant as it might initially appear to be. As I mentioned earlier, a lot of physicists suspect that the universe emerged from an uncaused quantum fluctuation. As Victor Stenger has argued, if this happened once, it might have happened many times, and in fact it probably did. Why would it not?[21]

Let's accept, then, that there may be multiple universes. How does this challenge the fine-tuning argument? The challenge comes from the possibility that, in different universes, there may be different values of the fundamental physical constants. In the vast majority of universes, the parameters would be unsuitable for the evolution of

[20] See, e.g., Rees (1997); Susskind (2006); Vilenkin (2006).
[21] Stenger (2007).

life. However, if there are enough universes, then in a small subset of universes, and purely by chance, the parameters would be suitable. The suitability of this particular universe would no longer be miraculous; if universes are sufficiently numerous, it would be miraculous if some *weren't* suitable for life. How did we get lucky enough to end up in one of the ultra-rare suitable ones? We come back to the anthropic principle: for us to be here asking this question, this universe *has* to be one of the lucky few.

If the multiverse scenario is valid, then marvelling at the fact that the universe is suitable for the evolution of life is a little like marvelling at the fact that someone beat the odds and won the lottery, while steadfastly ignoring the fact that millions of others played but did not win. If you only focus on the winner, it's very tempting to conclude that God must have somehow orchestrated the win – especially if the winner happens to be you. 'Sure, *someone* has to win', you might think. 'But why *me*?' The problem is that if you invoke God to answer this question, then instead of having to get your head around the fact that you were the one person in a million who won by random chance, you have to get your head around the fact that you were the one person in a million whom God selected to be the winner. How is that any better?

One version of the multiverse hypothesis employs Darwinian principles to explain the apparent fine-tuning of the universe. The physicist Lee Smolin has proposed that universes literally evolve.[22] It sounds crazy, but if you suspend your disbelief for a moment, I'll try to persuade you that it's not. To count as an evolutionary process, notes Smolin, three elements are needed: reproduction, heredity, and differential fitness (see Chapter 2). Applied to universes, this would imply the following. First, universes would have to have some means of giving rise to other universes (reproduction). Second, the physical parameters of these 'offspring' universes would have to be very similar – though not identical – to those of the 'parent' universe (heredity). And third, some

[22] Smolin (1992, 1997).

universes would have to produce more offspring than others – that is, universes with some sets of parameters would have to be fitter or more fecund than those with other sets (differential fitness).

Weird though it might sound, these conditions are not out of the question. We'll start with the issue of how universes might be able to reproduce themselves. The answer is tied up with the phenomenon of black holes. Black holes form when large stars explode as supernovae, and then the remnants of those stars collapse under the weight of their own gravity into an infinitesimal, massively dense point. The gravitational field of a black hole is so great that nothing can escape from it, not even light. A number of physicists have suggested that the formation of a black hole in one universe gives rise to the Big Bang of a new one. Thus our own universe may be the product of a black hole in another. If the fundamental constants of any given universe are such that it contains many large stars, that universe will eventually come to contain many black holes and will produce many offspring universes (differential fitness). If these offspring universes resemble the parents sufficiently (heredity), they too will produce many offspring universes. These 'fertile' universes will come to dominate the multiverse, while the less fertile ones take a backseat. Thus we have a genuine evolutionary process. If this theory is correct, our universe was 'designed' by natural selection to produce large stars and black holes. It was not designed to produce humans, life, or mind. These things are merely by-products of the process of cosmological evolution.[23]

---

[23] A legitimate question is what got the whole process up and running in the first place. Even granting that cosmological evolution is taking place now, we would still need an initial universe with at least one black hole to get the process started. If we can account for the initial universe without reference to cosmological evolution (e.g., by explaining it as the result of an uncaused quantum fluctuation), then why can we not just explain our own universe in the same way? Cosmological evolution seems redundant. Smolin disputes this, suggesting that his theory helps to explain a number of otherwise inexplicable facts, such as why life appeared so quickly in the universe's history (i.e., pretty much as soon as it could). At this stage, though, the jury is still out.

Smolin's theory sounds fantastical at first, and not surprisingly it is highly controversial. For present purposes, though, this doesn't matter. A theory can be philosophically important even if it turns out to be false. The fact that cosmological evolution is possible in principle implies that God is not the only explanation for the (relative) bio-friendliness of this universe. Thus the bio-friendliness of the universe does not commend belief in God as the only rational option. Just as the design found in organisms doesn't imply a designer (intelligent or otherwise), neither does the fine-tuning of the laws of nature imply a fine-tuner (divine or otherwise). The fine-tuning argument fails to establish its conclusion: that it is reasonable to believe in God. Once again, theists should ask themselves: what reason *do* I have to believe?

Mind the gap

I began this chapter with the suggestion that there is no role for God within the evolutionary process. We've now looked outside that process and found that there's also no role for God in kick-starting life, or in bringing into existence a universe in which life can evolve. But there's still another tack we could take. It might be argued that, up until this point, we've restricted our gaze to the physical universe. But what about the so-called non-physical realm, the realm of mind? A lot people's view on the subject of evolution can be summed up in four words: 'Body yes, mind no.' Ever since Darwin, some have accepted that the body is a product of evolution, but denied this of human consciousness, or of particular mental faculties such as morality or the aesthetic sense.[24] A number of Catholic thinkers, for instance, have suggested that evolution took place in largely the way that scientists tell us it did, but that at some point in the process God chose a pair of hominids and infused in them an immortal soul,

[24] See, e.g., Polkinghorne (1994); Roughgarden (2006); Wallace (1870).

fully equipped with free will and a knowledge of good and evil. This moment was the true origin of our species, the point at which God created the first humans from an evolving lineage of soulless matter. Since then, God has created each human soul directly and immediately.[25]

Alfred Wallace, co-discoverer of the principle of natural selection, was among the first to deny that the human brain or mind is a product of evolutionary processes, and to argue instead that it must be the handiwork of a divine creator. Originally, he had thought that mind could be explained in naturalistic terms, but later he changed his tune (much to Darwin's chagrin). Wallace's main argument was that we possess many mental abilities that provide no reproductive advantage and that therefore could not have evolved. His list of unevolvable traits included language, mathematics, abstract reasoning, music, the aesthetic sense, and the capacity for morality. And not only do these abilities lack any evolutionary function, they can be found even in peoples who do not use them. Wallace pointed out that 'savages' do not compose sonatas, play chess, or do calculus, but they have the capacity to do all these things. He put the point like this:

> A brain slightly larger than that of the gorilla would, according to the evidence before us, fully have sufficed for the limited mental development of the savage; and we must therefore admit, that the large brain he actually possesses could never have been solely developed by any of those laws of evolution, whose essence is, that they lead to a degree of organization exactly proportionate to the wants of each species, never beyond those wants.
>
> Natural selection could only have endowed the savage with a brain a few degrees superior to that of an ape, whereas he actually possesses one very little inferior to that of a philosopher.[26]

These words sound racist to the modern ear – and they are – but it's only fair to point out that Wallace was considerably *less* racist than

[25] See, e.g., Lewis (1940); Maritain (1997); Mivart (1871).
[26] Wallace (1870), pp. 343, 356.

most of his contemporaries in his belief that so-called primitive peoples were endowed with a comparable level of intelligence to Europeans. In any case, the real question is: if our mental powers could not have evolved, how did we come to possess them? Wallace came up with a very interesting, utterly implausible answer to this question. He proposed that these traits had been selectively bred into us by a superior intelligence.

> The inference I would draw from this class of phenomena is, that a superior intelligence has guided the development of man in a definite direction, and for a special purpose, just as man guides the development of many animal and vegetable forms.[27]

In other words, intelligence, according to Wallace, is a product not of natural selection but of *super*natural selection. Although no one has taken up this particular aspect of his position, the argument that certain aspects of mind are unevolvable, and must therefore be the result of divine intervention, has been a regular feature of the intellectual landscape ever since Wallace made it. It was echoed, for example, in a recent argument made by John Polkinghorne, a theoretical physicist who resigned his professorship at Cambridge University in the late 1970s to become an Anglican priest. Polkinghorne argues that, unaided, natural selection could only endow our minds with aptitudes necessary for coping with the adaptive challenges of everyday life. Are those the only aptitudes we possess? Absolutely not! Polkinghorne is particularly impressed by the fact that the human mind is able to comprehend the abstract truths of mathematics, quantum physics, and relativity theory – truths that have no bearing whatsoever on our survival prospects or our likelihood of passing on our genes. These abilities prove, he argues, that our minds are not products of natural selection. They provide compelling evidence for the existence of a designing and loving God.[28]

[27] Wallace (1870), p. 359.
[28] Polkinghorne (1994).

Time to evaluate the arguments. First, it needs to be pointed out that the idea that the body is a product of evolution, but that the mind or soul is not, goes against the spirit (so to speak) of Darwinism. It posits a radical discontinuity in the mode of origination of two parts of the human organism, and a radical discontinuity between human minds and the minds of other animals. Although such discontinuities are not logically inconsistent with evolutionary theory, they must be viewed with a great deal of suspicion after Darwin. If we wish to maintain that the human mind is exempt from the process of natural selection, we must provide strong grounds for this otherwise inexplicable exception. The fact that one wishes to square evolution with one's religious beliefs is not enough.

So how strong are the arguments? Well they're certainly not the weakest on offer, but they're unlikely to impress anyone who's not already a believer. According to modern evolutionary psychologists, the basic design of the mind, like any other functionally complex aspect of the organism, is the product of a gradual accumulation of variants that enhanced the fitness of their bearers. Convincing evolutionary rationales have been put forward for many of the abilities that Wallace and others claimed could have no evolutionary function. This includes the aesthetic sense, the capacity to acquire language, and basic mathematical competency.[29] There have also been major breakthroughs in unravelling the evolutionary foundations of human morality, which we'll examine in Chapter 11. The mere fact that it is possible to concoct Darwinian explanations plausible enough to attract serious consideration undermines the suggestion that the existence of these mental capacities compels belief in an intelligent designer (or designers – let's not forget Hume's critique of the argument from design). The theist's arguments lose all their bite.

[29] On the aesthetic sense, see Thornhill (1998). On language, see Deacon (1997); Pinker (1994); Pinker and Bloom (1992). On our basic facility with numbers, see Butterworth (1999); Dehaene (1997); Geary (1995).

Still, there's no getting around the fact that the human mind can do lots of things that don't seem to have anything to do with survival or reproduction. As Polkinghorne noted, this includes the ability to comprehend advanced mathematics, quantum physics, and relativity theory. Must we conclude, then, that the mind is a gift from God? No. It's quite clear that we can train our brains (that is, our brains can train themselves) to do things they were not specifically designed to do. The human brain was not specifically designed to play guitars or memorize football scores, for example, but we can use it for those purposes. In exactly the same way, the human brain was not specifically designed to understand complex mathematics, quantum physics, or relativity theory, but we can understand these things by co-opting psychological competencies that were selected for other purposes. The fact that people can do this is no more mysterious – and no better proof of God's existence – than the fact that people can learn to ride bicycles or that screwdrivers can be used as letter openers.

While we're on the topic of relativity theory and quantum mechanics, we should note that not everyone can comprehend these twin pillars of modern science. Does this mean that not everyone's mind is a gift from God? And why, if reason really is a gift from God, do we not have a *better* understanding of relativity and quantum physics? Why are there these areas of knowledge that reason is almost incapable of grasping? Quantum physics is so counterintuitive that even its practitioners claim not to truly understand it. This is hard to explain on the assumption that God created our mental faculties with the specific intention that we be able to comprehend the fundamental nature of reality. In contrast, an evolutionary explanation can easily account for the data. The human mind evolved to comprehend the human-scale world, the world of moveable objects and walkable distances. When we use it to do something utterly different, such as trying to comprehend the subatomic world (or the behaviour of matter in a strong gravitational field, or the behaviour of matter approaching the speed of light, etc.), we're able to do so only clumsily and imperfectly. Our incomprehension in the face of these distant domains of reality is exactly what we'd expect if our minds were a product of natural selection.

The problem of consciousness

Even if we can explain all the capacities of the human mind in naturalistic terms, there still remains a problem. If these capacities are no more than the workings of an evolved brain, how can we account for the existence of consciousness? If human beings are no more than matter in motion, how can we be aware of the smell of a rose, the taste of an orange, the sight of a deep red sunset? How can mere matter be aware of anything at all? Wallace argued that it could not, and thus that we must be more than mere matter.

> If a material element, or a combination of a thousand material elements in a molecule, are alike unconscious, it is impossible for us to believe, that the mere addition of one, two, or a thousand other material elements to form a more complex molecule, could in any way tend to produce a self-conscious existence. The things are radically different.[30]

Evolutionary theory can explain how clumps of matter come to be organized in such a way that they can take in information about the surrounding world and use it to guide their behaviour. But it cannot answer the deeper question of why these information-processing devices are conscious. It cannot explain why matter apparently has the capacity to become conscious when it is organized and functioning in a particular way. As the theologian Richard Swinburne asked, if natural selection is our sole author, why are we not just insentient robots?[31] For many people, it is consciousness that proves that there must be more to us than matter. For many people, the only way to explain the non-physical component is to invoke God.

[30] Wallace (1870), p. 365.
[31] Swinburne (1997).

Explaining consciousness is certainly difficult and maybe impossible.[32] But we have to ask two questions. First, does positing God do anything to solve the problem? And, second, even if we cannot explain consciousness, can we nonetheless find reasons to accept that it is a natural phenomenon? Unfortunately for the theist, the answer to the first question is 'no', and the answer to the second is 'yes'. Regarding the first question – whether positing God helps to dissolve the mystery of consciousness – we can start by asking: why exactly would God want to imprison immaterial souls in these strange animated hunks of evolved matter, hunks of matter that would otherwise be as lacking in consciousness as mountains or mud? It seems a little odd, especially when you think through some of the implications. Presumably God designed the basic nature of the soul so that its desires and ambitions would coincide with the evolved survival needs and behaviours of these aggregations of matter. How is this any less peculiar than God imprisoning souls in, say, snow-flakes, and bequeathing them a psychology suitable for that station in life – a desire, perhaps, to form into unique crystalline shapes and float gently to the ground? Another question for the theist: if God created consciousness, where did God's consciousness come from? God *is* conscious, right? If so, then to explain human consciousness in terms of God's conscious creative activity is to beg the question at issue. In other words, it is to assume the existence of the very thing that one is trying to explain: consciousness. We might not have a complete understanding of consciousness (or of anything else in the world), but there's no reason to think that the explanation for the residue will be 'it's magic' – which is what the theistic explanation amounts to. Where there are gaps in our knowledge, we should admit our ignorance, not resort to the unsupported answers of ancient peoples who were even more ignorant than we are.

---

[32] For attempted evolutionary explanations of consciousness, see Dennett (1991); Humphrey (2000, 2006).

Regarding the second question – whether we can find reasons to accept that consciousness is a natural phenomenon, even in the absence of a complete understanding of how this could be so – we must start by conceding that it is hard to imagine how mere matter could be capable of consciousness. But simply positing another, non-natural 'substance' (mind or soul) that *is* capable of consciousness doesn't solve the problem; it just relocates it. If it seems unproblematic that some immaterial substance could be capable of consciousness, whereas matter could not, this is only because we implicitly define this substance as capable of consciousness. And if we can simply assert that an immaterial mind-substance is capable of consciousness, why can we not simply assert this of matter? The latter option is actually preferable, because it sidesteps the problem of explaining how something immaterial could interact with the material world (e.g., how a non-physical mind could direct the behaviour of a physical body). So here we have a strong reason to favour the view that consciousness is a natural phenomenon.

It seems, then, that we must accept that the mind is the activity of an evolved brain, rather than some kind of non-physical entity that interacts with the brain. As soon as we do accept this, though, something very interesting happens. We began with the argument that the existence of the mind stands as evidence of God. However, if the mind is the activity of a physical organ – the brain – it actually does the opposite: it stands as evidence *against* the existence of God. For one thing, the dependence of mind on brain implies that the mind can only survive for as long as the brain does, and thus that there can be no life after death. (I'll say more about this in Chapter 8.) An integral part of most people's conception of God is the idea that God grants us continued existence beyond the grave. As such, the fact that the mind is the activity of the brain is inconsistent with the God of most believers. Some claim that natural selection was God's way of creating minds. However, they tend to gloss over the fact that this makes the mind mortal and that it therefore makes God much less plausible.

It gets worse. As soon as we view the mind as an adaptation, a mechanism designed to perpetuate the genes that give rise to it, the idea that God is anything even vaguely akin to a human mind loses any traction it might once have had. As an editorial in *Nature* put it:

> The suggestion that any entity capable of creating the Universe has a mind encumbered with the same emotional structures and perceptual framework as that of an upright ape adapted to living in small, intensely social peer-groups on the African savannah seems a priori unlikely.[33]

The mind is pretty transparently a device designed to help the organism survive and reproduce and aid kin. This makes sense in the case of human and other animal minds; we are products of an evolutionary process that assembles these kinds of devices. Why, though, would an uncreated, all-powerful God resemble such a thing?

An evolutionary perspective on the mind also undercuts some of the intuitions that make God seem so eminently plausible. Some commentators, such as the physicist and science writer Paul Davies,[34] argue that the universe is too well ordered to be a mere mindless accident, and therefore that mind or intelligence must underlie the order we see around us. However, viewed from an evolutionary perspective, mind is not the *cause* of the order in nature; mind is an *example* of the order in nature – something to be explained rather than the explanation for everything else. Once we view the mind as an adaptation, the suggestion that everything must ultimately trace back to mind, or intelligent agency, starts to look about as credible and reasonable as the suggestion that everything must ultimately trace back to eyes or wings or bacterial flagella. At first glance, mind and consciousness seem to hint at the existence of a creator. On closer inspection, they do not. They make God even less plausible than he was before.

---

[33] *Nature* (2007), p. 753.
[34] Davies (1992).

Conclusion

In this chapter, we have considered three attempts to find a role for God outside the evolutionary process: conjuring life out of inanimate nature, conjuring a bio-friendly universe out of nothing, and injecting mind or consciousness into a world of mere matter in motion. Each attempt has come up short, and evolutionary theory contributes to this conclusion in every case. There's a more general point to make as well. Throughout the chapter, we've been dealing with an old theological habit: locating God wherever a gap is found in current scientific explanations. Such arguments are known as *God-of-the-Gaps* arguments. Using gaps as proof of God is an inherently risky strategy. First, even if there really is a legitimate gap (which often there is not), we have no reason to think that God is the appropriate gap filler. Second, science has a nasty habit of filling gaps that once seemed unfillable. As such, gap arguments expose the theist to the possibility of an embarrassing disproof, and many theologians prefer not to risk making them. Instead, they look for other ways to integrate God into their view of the world. For example, according to the nineteenth-century evangelist Henry Drummond (who actually coined the term 'God of the Gaps'), we should view *all* of nature as God's work, rather than confining God to gaps in the scientific account of the universe. The problem is, though, that when we consider 'God's work' through the lens of evolution, we come face-to-face with a new version of a very old problem: if there is a God, why is there so much suffering in the world? That's the issue we'll look at next.

SIX

Darwin and the problem of evil

Whatever the God implied by evolutionary theory and the data of natural selection may be like, he is not the Protestant God of waste not, want not. He is also not the loving God who cares about his productions ... The God of the Galapagos is careless, wasteful, indifferent, almost diabolical. He is certainly not the sort of God to whom anyone would be inclined to pray.

David Hull (1991), p. 486

If a tenth part of the pains which have been expended in finding benevolent adaptations in all nature, had been employed in collecting evidence to blacken the character of the Creator, what scope for comment would not have been found in the entire existence of the lower animals, divided, with scarcely an exception, into devourers and devoured, and a prey to a thousand ills from which they are denied the faculties necessary for protecting themselves!

John Stuart Mill (1874), p. 58

In the part of this universe that we know there is great injustice, and often the good suffer, and often the wicked prosper, and one hardly knows which of those is the more annoying.

Bertrand Russell (1957), p. 13

What's the problem?

Our main conclusion so far is that evolutionary theory dissolves some of the most potent reasons for believing in God. Before going any further, it's worth noting that, even if that's all it did, this would not be a trivial implication. After all, without a reason to believe something, belief is unreasonable. But in any case, evolutionary theory does a lot more than remove reasons for belief; it contributes positive reasons for disbelief. We've already touched on one of these: incompetent design in nature is evidence against certain conceptions of God, in particular the hands-on creator of the Creationists and some Theistic Evolutionists. However, there's another difficulty for the individual who accepts evolution but wishes to reconcile this with a belief in God, a difficulty that applies whether God directly intervened in the evolutionary process or simply set the ball rolling and then sat back and watched. This is a new variant of the age-old 'problem of evil' as an argument against the existence of God.[1] The problem can be summed up in a single, simple question: if God exists, and if God is good, why is there so much evil (i.e., suffering) in the world?

Evolutionary theory profoundly exacerbates the traditional problem of evil. The history of life on earth is a history of carnage and waste, a vast and bloody struggle for existence. And the evolutionary drama has been playing out not for thousands of years but for billions. In short, the problem of evil is much larger than anyone ever dreamed. Furthermore, an evolutionary perspective combats our native tendency to focus exclusively on members of our own species, and brings to our attention the suffering of non-human animals throughout the vast expanses of evolutionary time. If evil was a problem before Darwin, it is a thousand times the problem now.

[1] On the problem of evil, see Adams and Adams (1990); Howard-Snyder (1996).

Atheism's killer argument

The problem of evil has been haunting theists for thousands of years. The challenge is to reconcile the existence of evil and suffering with the fact that the supposed creator is supposedly all-knowing, all-powerful, and all-good. If God knows everything there is to know, he knows about all the pain and suffering taking place on the surface of this planet (and probably others). If God is completely good, he presumably wishes to eliminate this pain and suffering. And if God is omnipotent, it is within his power to do so. Nonetheless, pain and suffering continue to exist. Why? Why doesn't God intervene? Remember, this is a God who allegedly concerns himself with the fall of every sparrow. The problem is not just that God doesn't step in and help. If God could have created any possible universe, why did he create one in which there was evil in the first place? For that matter, why did he create *any* universe at all? He presumably didn't have to. But he did it nonetheless, and therefore he's responsible for what goes on in the world. Perhaps we can agree with Albert Einstein that only God's non-existence would excuse him.

The apparent clash between the nature of God and the existence of evil leaves the thoughtful theist with a number of options, none of which is particularly appealing. One is to concede that God is powerless to prevent evil. Another is to concede that God is not all-good. Maybe God is evil! Or maybe God is indifferent rather than evil. The suffering of evolved sentient life is important to us, but perhaps in the grand scheme of things it is utterly inconsequential. Perhaps we are merely God's playthings, and he's experimenting with us in the same nonchalant way that we might experiment with a computer simulation. A final option is that we were wrong about God all along – that we've been making a terrible mistake for the last few thousand years and that there is no God.

It's worth noting that not all atheists accept the merits of the argument from evil. Richard Dawkins is one who dissents, noting that the problem of evil 'is an argument only against the existence of a good God … for a more sophisticated believer in some kind of supernatural intelligence, it is childishly easy to overcome the problem of evil'.[2] Admittedly, the problem of evil works only for one conception of God – God as all-powerful, all-knowing, and all-good – and at first glance, it might seem an attractive option to jettison one or a few of these attributes. Most theists probably didn't have a clear idea that God was omnipotent or omniscient or omnibenevolent before thinking about the matter, so it might seem that they would have little to lose by accepting a diminished conception of the deity. Unfortunately, though, it's not so simple, because on closer inspection the implications of such a conception undermine many common religious practices and attitudes. If God is not omnipotent, how could you trust him to look after you or your nearest and dearest? If God is not omniscient, why should anyone accept his pronouncements without subjecting them to critical scrutiny and, wherever appropriate, empirical examination? In other words, why should we take anything on faith? Finally, if God is not good, why should we worship or obey him? Indeed, if God were evil, wouldn't we have a moral obligation *not* to obey him?[3] The problem of evil is not nearly as trivial as Dawkins suggests. It strikes at the very heart of theistic belief.

Darwin and evil

The problem of evil didn't play a big role in my own transition to atheism. But for many people it was the deciding factor, the straw that

---

[2] Dawkins (2006), p. 135.

[3] One reason we might continue to worship and obey God, even if God were evil, would be to avoid his wrath and obtain the good things that such a powerful being could proffer. But this is the same reason that someone might suck up to a despotic king or leader. If God existed, it might be *prudent* for us to obey his commands, just as it would be prudent to obey the commands of any powerful individual, especially one with the reputation of a despot. However, it would hardly be admirable. We would not be *morally* obliged to obey God unless we had some reason to think that God were good.

broke the camel's back. Not only that, but the specific contribution made by a Darwinian perspective has persuaded a lot of people to drop their faith. The following quotation provides a lucid description of the problem faced by the person who accepts that evolution took place, but wishes to retain a belief in God:

> Could an Almighty God of love have designed, foreseen, planned, and created a system whose law is a ruthless struggle for existence in an overcrowded world? Could an omnipotent, omniscient, and omnibenevolent God have devised such a cold-blooded competition of beast with beast, man with man, species with species, in which the clever, the cunning, and the cruel survive?[4]

You might be surprised to learn that the author of these words was a clergyman. You might be less surprised if I tell you that, by the time he put pen to paper and wrote this, he was actually a *former* clergyman, who had come to the conclusion that atheism was the only intellectually honest position available for anyone who truly understood evolutionary theory. Darwin himself was always deeply disturbed by the suffering that natural selection entailed. As I mentioned in Chapter 4, he occasionally entertained the possibility that some kind of God had designed the laws of nature. But he had great trouble imagining that the universe was guided or controlled by any kind of *benevolent* force. Shortly after the publication of the *Origin*, Darwin wrote to his friend, Asa Gray:

> I had no intention to write atheistically. But I own that I cannot see as plainly as others do, and as I should wish to do, evidence of design and beneficence on all sides of us. There seems to me too much misery in the world. I cannot persuade myself that a beneficent and omnipotent God would have designedly created the Ichneumonidae [a parasitic wasp] with the express intention of their [larvae] feeding within the living bodies of Caterpillars, or that a cat should play with mice.[5]

---

[4] Cited in Haught (2000), p. 21.
[5] Cited in F. Darwin (1887b), pp. 311–12.

Darwin is not the only evolutionist to have expressed such sentiments. Many have looked at natural selection and judged it to be morally unacceptable. George C. Williams, a prominent evolutionary biologist, went so far as to call it 'evil'.[6] Natural selection is a massively inefficient process. For every beneficial mutation, there are countless others that inflict gross and pointless suffering on their bearers; for every successful variant, thousands of organisms perish miserably. Not only that, but even the successful ones – i.e., organisms that have more offspring than their neighbours – often end their days in suffering. Dawkins put the point well:

> Nature is not interested one way or the other in suffering, unless it affects the survival of DNA. It is easy to imagine a gene that, say, tranquilizes gazelles when they are about to suffer a killing bite. Would such a gene be favored by natural selection? Not unless the act of tranquilizing a gazelle improved that gene's chances of being propagated into future generations. It is hard to see why this would be so, and we may therefore guess that gazelles suffer horrible pain and fear when they are pursued to death – as most of them eventually are. The total amount of suffering per year in the natural world is beyond all decent contemplation. During the minute it takes me to compose this sentence, thousands of animals are being eaten alive; others are running for their lives, whimpering with fear; others are being slowly devoured from within by rasping parasites; thousands of all kinds are dying of starvation, thirst and disease.[7]

As this passage suggests, not only is the process of natural selection ruthless, but so too are many of its products. Numerous unpleasant facts about the natural world can be pinned directly on natural selection. Among parental species, mothers usually abandon their weak or disabled offspring – they just let them die. Likewise, when a lion takes over a pride, it typically kills the existing cubs – those sired by rival males.[8] Infanticide, siblicide, and rape are common in the

[6] Williams (1993).
[7] Dawkins (1995), pp. 131–2.
[8] Schaller (1972).

animal kingdom. All these unpleasant phenomena make good sense in terms of selection for genes that get themselves copied into future generations at a faster rate than competing versions of the same genes (competing *alleles*). The question is: why would a kind and benevolent God choose such a reprehensible process as his means of creating life?

One might respond that the picture of nature that has emerged since Darwin's day is not one of relentless cruelty, and that the extent of the competition and selfishness in nature has been vastly overestimated.[9] Evolution doesn't only produce unpleasant traits; as we'll see in Chapter 11, parental care, altruism, and other morally desirable behaviours also have their origin in natural selection. But although nature might not be unrelentingly red in tooth and claw, there's still more than enough redness to give the theist pause. Furthermore, as George Williams noted:

> Attempts to demonstrate the benevolence of Nature often take the form of name changing. The killing of deer by mountain lions meant 'nature red in tooth and claw' to a generation of 'social Darwinists.' To a more recent generation it has become Nature's kindness in preventing deer from becoming so numerous that they die of starvation or disease ... The simple facts are that both predation and starvation are painful prospects for deer, and that the lion's lot is no more enviable.[10]

There's no avoiding this unpleasant fact: there has been pain and suffering on this planet for as long as there have been multicellular organisms with nervous systems capable of mocking up the experience of pain and suffering. Thus the planet has been host to at least half a billion years' worth of suffering. In Chapter 4, we saw that many theists claim that God chose natural selection as his means of creating life. But when you really think about what this implies, you recoil from the idea in horror. Why the bloodbath?

---

[9] See, e.g., de Waal (1996).
[10] Williams (1966), p. 255.

The question of consciousness in non-human animals

This entire line of argument rests on the assumption that at least some non-human animals are sentient beings, capable of experiencing pain and other unpleasant psychological states. It is open to the theist to deny this. The celebrated seventeenth-century philosopher René Descartes argued that non-human animals are unfeeling machines, no more capable of suffering than wind-up toys. Few have been able to accept such a counterintuitive position, but if it were true it might provide a way of escaping the Darwin-enhanced problem of evil. If non-human animals were unconscious automata, such things as predation and infanticide would only be aesthetically unpleasant. On the other hand, to the degree that non-humans are sentient, the cruel products of natural selection represent precisely the kind of evil that conflicts with the existence of a loving God.

The idea that animals don't have consciousness suggests an uncomfortable parallel. When the early Spanish explorers first arrived in South America, they seriously debated whether the indigenous peoples of these lands had souls. Similarly, Christians in the past asserted that non-white people were soulless, in exactly the same sense that Descartes claimed this of non-human animals. Fortunately, most people are now free of this kind of bias. It is perfectly possible, though, that the denial of a 'soul' or consciousness in animals is an equivalent bias, a bias which future generations will look at with the same incredulity and contempt that we have for those who denied souls to non-whites. At the very least, this possibility should give pause to those of us who deny that any non-human animals are conscious. We're certainly capable of making mistakes along these lines.

It might be objected that when we attribute consciousness to other animals, we are simply anthropomorphizing – that is, we are attributing uniquely human traits to animals that don't possess them. And we should certainly be aware of this potential pitfall. We shouldn't forget, though, that there is also the danger of making the opposite mistake: viewing shared traits as uniquely human. Often this possibility is overlooked, perhaps in part because there is no simple word or phrase for it.[11] However, there is reason to suspect that the second mistake might be more common than anthropomorphism itself. People kill non-human animals for food, for their skins, and sometimes just for fun. We enslave animals and force them to work for us. We experiment on them and justify *their* suffering in terms of *our* advantage. Because most of us want to be able to view ourselves as good people (and, perhaps more importantly, because we want others to view us as good people), we may be motivated to view non-humans in such a way that these activities are rendered morally unproblematic. One way to do this is to view other animals as utterly different from us. Darwin wondered about this; he speculated that the reason we admonish each other to avoid anthropomorphizing, but not to avoid making the opposite mistake, is that we have an emotional need to rationalize our treatment of nonhumans.[12]

There are various reasons to think that other animals are conscious and can experience pain and suffering.[13] First, as we've already seen, an evolutionary perspective tells us that the mind is the activity of the brain. If human brains are conscious, why would this not be the case for non-human brains, especially those that most closely resemble our own? Second, we have largely the same evidence that other animals experience pain as we do that other *humans* experience pain: we infer it from their behaviour. If the evidence is good enough in the human case, why isn't it in the case of non-humans? I'm not going to deny that it's possible in principle that every animal bar *Homo sapiens* lacks consciousness – but it is also possible in principle that every human being bar oneself lacks consciousness. What reason do we have to take either possibility seriously?

[11] My suggestion: denying shared traits to animals that in fact possess them is *deanimalizing* them.

[12] Barrett and Gruber (1980).

[13] On the question of consciousness in other animals, see Marion Stamp Dawkins (1993).

There's one more thing to consider. Some argue that we should not assume that other animals have consciousness until we have indisputable evidence that this is the case. I hold the opposite view: we should not assume that other animals do *not* have consciousness until we have indisputable evidence of *that*. If we assume that they're conscious and we're wrong, the price for our error is trivial: we'll treat them more humanely than we need to. But if we assume they are not conscious and we're wrong about *this*, the price will be that we treat them much worse than we should. Either assumption may be false, but the first has far less horrible consequences than the second.

It seems, then, that we must conclude that the notion that at least some non-human animals are sentient and capable of suffering is an entirely reasonable one. Indeed, in the absence of a very strong argument to the contrary, it is unreasonable to think otherwise. There may be grounds to assume that humans have a *greater* capacity to suffer than any other animal. However, even if non-humans don't suffer as much as we do, haven't they suffered enough to call into question the existence of a benevolent creator and keep thoughtful theists awake at night?

Getting God off the hook

The apparent 'cruelty' of natural selection and its products does not *entail* the conclusion that there is no God. As Darwin pointed out, though, it is difficult to reconcile the suffering caused by natural selection with a belief in an infinitely good and omnipotent deity. We know it's difficult because there have been numerous attempts to do so, and none has been particularly successful. Any attempt to answer the problem of evil without giving up God is called a *theodicy*.[14] In the following pages, we'll take a whistle-stop tour of nine common theodicies, and then we'll zero in on three particularly

---

[14] Weisberger (2007).

important ones: the free will defence; the idea that the universe is a school for developing souls; and the idea that evil is an inevitable by-product of natural law (including the 'law' of natural selection). We'll see that all the standard theodicies fail dismally when confronted with the Darwinian version of the problem.

1. *Evil is not real. It is an illusion. We think that there's evil in the world but we're wrong about this*

It's almost impossible to see how this glass-half-full take on evil could work. Even if evil is an illusion, it's an unpleasant one (an evil one, one might argue). On top of that, it's not at all obvious what it could mean to say that a subjective state, such as pain or anguish, is only an illusion. Like any subjective state, suffering exists to the extent that it is experienced. If it's experienced, it's real. Even psychosomatic pain is real pain; it's just that it's caused by psychological factors rather than by organic damage. In what sense could suffering be an illusion?

2. *Evil is merely the absence of good, just as a shadow is merely the absence of light*

This sounds appealing, but that may be its only real virtue. If we consider pain at a neurological level, we see that it is not simply the absence of the activation of parts of the brain involved in pleasure or happiness. It involves the activation of areas of the brain that evolved specifically for the purpose of producing pain. The claim that evil is the absence of good, like the claim that evil is mere illusion, smacks of wishful thinking.

3. *Good can only be appreciated in relation to evil. Misery leads to a greater appreciation of happiness, and ugliness to a greater appreciation of beauty*

One way to start evaluating this idea is to ask: would it be possible to experience pleasure if you were not capable of experiencing pain? And the answer to this question is 'yes, it would'. The brain circuits underlying the experience of pleasure are largely independent of those underlying the experience of pain. Indeed, some people are born without the capacity to feel pain, and yet are still able to experience pleasure. Furthermore, even if this weren't the case, it would presumably be possible in principle to wire up a brain capable of appreciating beauty but not ugliness, pleasure but not pain, happiness but not misery. One could still argue that God created evil to *show up* the good, to really highlight and underline it. Unfortunately for the theist, one could just as well argue that God is evil and created the good to show up the evil – to really rub our faces in it. Both views are equally consistent with the evidence, and thus neither should be favoured over the other.

4. *Evil is a divine punishment for our sins, our crimes and misdemeanours. Natural disasters are warnings from God that we should obey his rules*

The main objection to this idea is that babies and non-human animals sometimes suffer horribly, despite the fact that they are not moral agents and are therefore incapable of 'sin'. As for the claim that natural disasters are divinely sanctioned warnings, wouldn't God be better off performing the occasional unambiguous miracle, rather than lashing out in a manner that could just as easily be explained in naturalistic terms?

5. *Evil is the work of Satan; God is not responsible*

Whereas some claim that evil is a punishment for our sins, others are less inclined to lay the blame so close to home, and attribute it to the Devil instead. But why can an omnipotent God not stop Satan in his hoofed tracks? One *could* take the view of the Zoroastrians: deny that God is omnipotent and claim instead that there is a genuine struggle between God and Satan. Most modern theists, however, hold that God is supreme, rather than the equal of Satan, the yin to Satan's yang. Therefore, this theodicy fails. God *could* stop Satan if he chose to. Why does he not?

6. *The rewards of heaven are so great that they outweigh the suffering of earthly life*

First point: what about non-human animals? Has every non-human animal that suffered throughout evolutionary history received divine compensation? Is there divine compensation for all the gorillas slaughtered so their hands could be used as ashtrays, or all the tigers slaughtered so their penises could be used as aphrodisiacs that don't work? Is there divine compensation for all the billions of animals raised and slaughtered in factory farms each year? Second point: even if the answer to all these questions were 'yes', how would that excuse or explain evil, or reconcile it with the existence of an all-good, all-powerful God? As George Smith pointed out in his excellent book, *Atheism: The Case against God*, if a father viciously beat his son, this injustice would not be erased if the father then gave the son a lollipop, or even a thousand lollipops. He would still have done something terribly wrong.[15] Ditto God. Furthermore, given that the mind is dependent on the brain, this suggests that the mind could not survive the death of the body, and thus that there could be no heaven for us to enjoy after death. Thus, if there is a God, he has created beings that can suffer but that cannot reap a compensatory eternal reward afterwards. The problem of evil remains.

7. *God's hands were tied; he had no choice but to create the world in the way that he did*

According to proponents of this view, it was simply not possible to create a world in which good existed but evil did not. This is the only logically possible world, or alternatively, as Leibniz proclaimed, it is the *best* of all possible worlds: the world with the optimal balance of good to evil. (Try proving *that* wrong!) Again, this theodicy is ultimately unconvincing. It may sound plausible only because we lack the imagination to think of alternative worlds. Presumably that

---

[15] Smith (1980).

is not a problem that would afflict God. Like many theodicies, this one underestimates omnipotence. Furthermore, even if evil *is* inevitable in every possible universe, one could still ask why God elected to create the universe to begin with.

> 8.   *Pain and suffering stem from the operation of an evolved warning system without which we could not survive. Although these states are far from fun, they guide people and other animals away from sources of potential harm*

The main problem with this theodicy is that not all pain serves as a warning. Much of it is a pointless by-product of a warning system that does not know when warnings are useless.[16] For a deer dying slowly in a forest fire, frightened and alone, there is no longer any adaptive benefit in experiencing pain; nonetheless, the deer's pain system continues to function perfectly till the end. The pain is gratuitous, and it is gratuitous pain and suffering that most persuasively challenge the existence of God.

> 9.   *The ends justify the means. Despite all the unimaginable suffering it has caused, natural selection is justified because it ultimately culminated in human beings, a species approaching the likeness of God, capable of understanding and worshipping the deity, and capable of freely choosing virtuous actions*

This is a common theodicy among people trying to square evolution with the existence of a benevolent creator. Unfortunately, it's also another dead end. If the evolutionary process was designed to produce our species, the price was the sacrifice of the majority of life forms that have ever existed. How could the death and suffering of millions of species be justified in terms of the good of just one? Surely an omnipotent God could find another, nicer way to create us. Besides, what makes us think we're so great anyway? This is Bertrand Russell's mischievous commentary on this whole line of thought:

---

[16] Rachels (1990).

Since evolution became fashionable, the glorification of Man has taken a new form. We are told that evolution has been guided by one great Purpose: through the millions of years when there were only slime, or trilobites, throughout the ages of dinosaurs and giant ferns, of bees and wild flowers, God was preparing the Great Climax. At last, in the fullness of time, He produced Man, including such specimens as Nero and Caligula, Hitler and Mussolini, whose transcendent glory justified the long painful process.[17]

This brings to a close our whistle-stop tour. Now we'll take a longer look at some of the most important theodicies, starting with the free will defence.

Freedom and evil

The most popular response to the problem of evil is the free will defence.[18] The basic idea is that God gave us freedom of will, and that evil arises from the sinful abuse of this unique faculty. Evil stems from our choices, not from God's, and for that reason it is we, and not God, who are responsible for evil. Of course, if God hadn't given us free will in the first place, committing evil would not be an option. But according to advocates of this theodicy, the value of free will is so great that it outweighed the risk that it would be abused by the Neros and Hitlers of this world. That being the case, God acted in good conscience in creating creatures capable of choosing evil.

It might be expected that evolutionary theory would undermine the free will defence by undermining free will itself. And, admittedly, the notion of free will does start to look a bit fishy when viewed through an evolutionary lens. When did we evolve free will? Did *Homo habilis* have it? Or was *Homo erectus* the first free hominid? These questions make the whole concept of free will seem a little silly. The fact is, though, that an evolutionary perspective doesn't make

---

[17] Russell (1950), p. 84.
[18] See, e.g., Plantinga (1974); Swinburne (1998).

free will any sillier than it was before. If free will were possible in principle, one could maybe argue that it evolved, slowly and by degree, in the same way that consciousness did.[19] But this is a moot point because free will is not possible in principle. The concept is incoherent.[20] Consider this: our decisions and actions are presumably either caused or they're random. To paraphrase David Hume, if they're caused, we don't have free will, whereas if they're random – well again, we don't have free will. We choose our own actions, sure. However, our choices, just like everything else in the universe, are caused. And if they're not, if they're just random, what kind of free will would that be?

Even if free will *were* possible in principle, and even if God had gifted us this mysterious faculty at some point in our evolutionary past, it's not clear that this would remove God from responsibility for the evil it made possible. If God created human beings knowing that there was even the remotest chance that they might do the sort of evil things they do (or worse), then surely he was acting recklessly and irresponsibly. Just as Dr Frankenstein was ultimately responsible for the harm caused by his monster, so too God would be ultimately responsible for the harm caused by his botched creations. If God were one of us, he'd surely be charged with gross negligence. And one of the first questions his prosecutors would ask would be: 'Why didn't you intervene to stop people using their free will to do evil?' We all agree that freedom is good, but virtually everyone agrees that it's OK to prevent people raping or murdering others. Indeed, most people think it would be wrong *not* to do so if they could. Surely we can apply

---

[19] For an attempted evolutionary explanation of free will, see Dennett (2003).

[20] See, e.g., Radcliffe Richards (2000); Strawson (1986). To be precise, this is the case for one particular conception of free will: libertarian free will. This is the view that our actions or decisions are not completely determined by prior causes, and that human beings have ultimate responsibility for what we do. It is reasonable to focus on this conception of free will because, first, it represents the common-sense view on the topic, and second, the free will defence assumes this conception.

this standard to God as well as to ourselves. Of course, there are those who would argue that God is right not to intervene. But if anyone really believed this, wouldn't they be questioning the morality of our own efforts to prevent evil? Wouldn't they assume that we should emulate God in this matter?

One more point before we move on. The free will defence might appear reasonable if we maintain a myopic focus on humans and humans alone. As we've seen, though, evolutionary theory forces us to take a wider perspective. And the free will defence doesn't even begin to account for the suffering of non-human animals. It is simply not applicable to the Darwinian version of the problem of evil – not unless we wish to argue that when an animal abandons one of her young, or when a lion or baboon slaughters another male's children, this is an abuse of free will. This is the most important deficiency of the free will defence. Most of the evil that is the focus of the Darwinian problem of evil is not due to the exercise of free will by moral beings, because most animals are not moral beings and most of the evolution of life on earth took place before the first moral beings evolved. The free will defence is inapplicable to most of the world's evil.

Whatever doesn't kill me ...

Most people think there should be less suffering in the world. This is the assumption we've been making thus far, and it's a natural enough assumption to make. Perhaps, though, we're wrong about this. Some commentators suggest that evil and suffering are as important to the soul as food and light are to the body. They are necessary ingredients for building character, fostering spiritual growth, and leading us to commit ourselves voluntarily to God. According to advocates of this view, such as John Hick and Richard Swinburne, God and evil seem hard to reconcile only because we're operating under the misapprehension that the world should be a hedonistic paradise. But we're not here to enjoy ourselves. Our true purpose is to transcend our selfish natures and become loving, generous, and virtuous souls. By struggling with evil and hardship, we develop maturity, moral character, and a relationship with God. The existence of suffering allows us to grow into moral citizens of God's universe. This applies not only to our own suffering but to the suffering of the people in our lives, which can prompt in us the development of sympathy and other virtuous traits.

The most influential variation on this theme is Hick's *soul-making theodicy*.[21] An important part of Hick's story is the idea that God deliberately left some room for doubt about his existence.[22] The reason for this apparently perverse decision was that our choice to come to him and commit to him would then be a truly voluntary one. If God's existence were entirely certain, no sane person would have any choice but to give themselves to God. To develop freely into a morally upstanding person, suggests Hick, it is necessary to live in a world in which it appears that maybe God does not exist – a world, in other words, with widespread hardship, suffering, and evil.[23] It would perhaps be unkind to suggest that one of the ways that God makes his existence uncertain is to create life through the process of natural selection, a process that does not require his help and that belies his reputation as a Good Guy. But, in any case, the core of Hick's theodicy is the idea that our true purpose in life is not to enjoy ourselves but to enter voluntarily into a personal relationship with God. We need trials and tribulations, tragedies and travails, slings and arrows, to push us towards this (entirely voluntary) decision.

Full marks for imaginativeness! My first reaction to this theodicy is to think that, if the argument from evil leads to such extreme and

---

[21] Hick (2007).

[22] See also Swinburne (2004).

[23] Schellenberg (1993) and Drange (1998) make a very different argument. They suggest that a benevolent God would make his existence known to sincere seekers of the truth, and thus that the ambiguity concerning God's existence is evidence for his non-existence.

implausible positions, this can only be taken as a measure of the success of the argument. But let's put aside any reservations that Hick's theodicy is pure and untrammelled storytelling, and see what happens if we take the soul-making concept seriously. What happens is that we come face-to-face with some extremely uncomfortable questions. First, if life really is a school for developing souls, why is the difficulty of the lessons not standardized? Why are some lucky souls blessed with a surplus of suffering whereas others have woefully little with which to build their characters? Wouldn't a sincere advocate of the soul-making theodicy have to agree that those who suffer more are being unfairly advantaged? Wouldn't a sincere advocate have to agree that, if we want to be good people, we should let our friends and loved ones suffer as much as possible? To do otherwise would surely be ungodly, and would deprive them of important opportunities for spiritual growth.

We also need to ask just how successful the world is as a school for improving character and bringing people to God. Certainly, hard times and personal tragedies do sometimes increase people's religiosity.[24] But often they do precisely the reverse. After Darwin's daughter Annie died at the age of ten – a particularly agonizing event in his life – Darwin stopped going to church. Annie's death hastened the demise of his commitment to religion, and obliterated the last scraps of his faith in a benevolent creator.[25] More generally, some small measure of suffering may build character, but the extremes of suffering do not. They psychologically scar people, leave them bitter and depressed, and can inflict emotional wounds that never heal. If the

---

[24] See Zuckerman (2007). It is ironic that religion flourishes most reliably in the presence of exactly the kinds of evils that call it into question.

[25] It is telling that it often takes a *personal* tragedy to kill off people's faith; after all, Darwin was obviously aware that other people suffered equivalent tragedies. It's as if people reason that it's OK for God to let terrible things happen to other people, but that if he lets something terrible happen to *them* … well in that case, there could be no God!

earth is a school for developing souls, then God is a poor educator. And why create beings in need of remedial education anyway? C. S. Lewis wrote that 'God whispers to us in our pleasures, speaks in our conscience, but shouts in our pains: it is His megaphone to rouse a deaf world.'[26] Fair enough, but couldn't he have just designed creatures that weren't deaf?

Next we need to question the moral character of a God who would establish the kind of system we've been discussing. Think about the idea that the existence of widespread suffering is intended to develop traits such as sympathy or trust in God. What kind of God would allow millions of humans and other animals to suffer and die in order to promote the spiritual development of the lucky few? This theodicy does not dissolve the problem of evil; it simply posits an evil God. Furthermore, like other theodicies, it doesn't touch the problem of non-human animals. Darwin's statement on this issue still stands as one of the best.

> That there is much suffering in the world no one disputes. Some have attempted to explain this in reference to man by imagining that it serves for his moral improvement. But the number of men in the world is as nothing compared with that of all other sentient beings, and these often suffer greatly without any moral improvement. A being so powerful and so full of knowledge as a God who could create the universe, is to our finite minds omnipotent and omniscient, and it revolts our understanding to suppose that his benevolence is not unbounded, for what advantage can there be in the sufferings of millions of the lower animals throughout almost endless time?[27]

Finally, let's consider the idea that God has deliberately concealed himself as a way of promoting our spiritual growth and making the decision to commit to him a voluntary one. Most philosophers find this type of argument – cute though it is – utterly unpersuasive. An initial problem is that in claiming that God has concealed himself,

[26] Lewis (1940), p. 91.
[27] Darwin (2002), p. 52.

the theist is tacitly conceding that the evidence for God's existence is unsatisfactory. There's another problem as well; maybe it has occurred to you already. The idea that God plants evidence designed to make his existence seem unlikely is a suspiciously convenient way of writing off all such evidence. It involves transmuting evidence against God's existence into evidence *for* God's existence. A clever ploy! Finally, although the concealed-God theory is logically possible, we have absolutely no reason to suppose that it is actually true. Our overall conclusion, then, is that the soul-making theodicy fails.

Evil is a by-product of natural law

There is one more theodicy to consider. Some commentators deny that evolutionary theory worsens the problem of evil, and argue instead that it *solves* it.[28] The idea is that, within an evolutionary framework, evil is not a direct product of divine design; it is an unfortunate but inevitable by-product of natural law. Any world governed by strict natural laws, such as the laws of physics and the law of natural selection, will inevitably contain suffering. A God who creates through unbroken natural law is therefore not responsible for the often-unsavoury goings-on of nature. This argument – if it works – absolves God of any guilt associated with the cruelty of natural selection and its products. Why? Because the blame can no longer be laid at the creator's doorstep. Evil and suffering are only inconsistent with the interventionist God of the Creationists and Theistic Evolutionists.

Once again, it's a valiant effort but the argument doesn't stand up to scrutiny. The first thing to notice – a point we've considered before in another context – is that a God who creates through unbroken natural law is not the God of traditional theism. It is an updated version of the God of deism, a fact that will stick in the throats of all

---

[28] See, e.g., Ayala (2007); Hunter (2001); Ruse (2001).

who wish to believe in a God who answers prayers, performs miracles, or otherwise intervenes in the world. For those who have already given up such beliefs, this is no problem. For a lot of theists, though, it will be a bitter pill to swallow.

And even those who can swallow it are not out of the woods yet, because *deism does not escape the problem of evil*. Even if God's role was merely to set the process of natural selection in motion, and even if he did not personally involve himself with the details of evolutionary history, he would still be responsible for the evil in the world. The theist might object that even an all-powerful God cannot do the impossible and that, having decided to create a lawful universe, evil is an inevitable side effect. But imagine that a lawyer made an equivalent argument: 'My client cannot do the impossible, and having decided to detonate a bomb in a busy public area, massive casualties were an inevitable side effect.' Sure, but the client didn't have to detonate the bomb! Furthermore, we need to ask again: why can God not intervene? To say that he can't because he's established a universe that runs in strict accordance with natural law is only to say that he can't because he's decided not to. Even if suffering really is an inevitable side effect of creation through natural selection, God chose natural selection as his means of creating life. As such, he is responsible for its effects.

Mysterious ways of moving

A final option would be to resort to a plea from ignorance: God moves in mysterious ways and we mere mortals cannot understand his divine purposes. What appears to us to be unnecessary suffering is in fact part of a plan, and that plan is good. I have to say that, if I'd been God, I would have done things very differently. But maybe God had his reasons, reasons that mere reason cannot comprehend. Maybe God had excellent reasons for allowing more than 99 per cent of his creatures to go extinct; for creating life through a process that entails unending misery; and for creating a world that looks like 'one great Slaughter-house', as Darwin's grandfather Erasmus put it.[29] Whatever these reasons are, however, we in our mortal ignorance and with our puny little minds are not up to the task of comprehending them.

The claim that God moves in mysterious ways is a last resort for theists when they're backed into a corner. Most of the time, they seem to have no difficulty understanding God's plan or deciding what's good and what's not. It's only when the evidence goes against them that they start waxing lyrical about God's inscrutability. And such claims are never used consistently. After all, if God's mysteriousness implies that we cannot blame him for the evil in the world, then neither should we credit him with the good things. Furthermore, a plea from ignorance could just as well be used to argue for the existence of an *evil* God rather than a good one: 'Sure, there's a lot of good in the world. But the evil God moves in mysterious ways and the things we think are good are actually part of a deeper and thoroughly evil plan.' When the same argument can be deployed in support of diametrically opposed positions, it's safe to ignore the argument. The evil-God argument cancels out the good-God argument. As a reason to believe in God, a plea from ignorance is scraping the bottom of the barrel.

Conclusion

Some suggest that positing God helps to make sense of the world. But the universe makes *less* sense on the assumption that God exists than on the assumption that he does not. This is nowhere more apparent than in considering the problem of evil. In his famous essay *Nature*, the philosopher John Stuart Mill described nature as cruel and violent, and denied that it could be under the rule of a benign providence. Evolutionary theory is entirely consistent with this view.

[29] E. Darwin (1803), line 66.

Although the problem of evil had haunted believers for thousands of years before Darwin, Darwin showed us just how serious the problem really is. Evolutionary theory prises our minds away from their usual narrow preoccupation with one species and one alone: *Homo sapiens*. It raises the salience of animal suffering, showing that the life of animals is often nasty, brutish, and short, and that nature is often red in tooth and claw. It also implies that the process that gave us the 'gift' of life involved the suffering of millions upon millions of animals. In short, the problem of evil tells us that, if there is a God, the Creationists should be right. A kind and benevolent God would have created life through a series of special creations, rather than through the long, painful evolutionary process. But as we saw in Chapter 2, the evidence is unequivocal: if there's a God, he opted for evolution. This strongly suggests that there is no God.

Nonetheless, like lawyers determined to exonerate a client caught red-handed, legions of thinkers have attempted to explain how a good God could create a world containing so much evil. Their efforts amount to little. At best, the theodicies show that the coexistence of God and evil is logically possible. In other words, the statement 'evil exists and so does God' is a statement like 'the earth is flat' (logically possible), rather than a statement like 'the earth is 100 per cent flat *and* 100 per cent spherical at the same time' (logically impossible). But this is all the theodicies show, and it's a paltry achievement. After all, it's also logically possible that evil exists but that God does not. These possibilities cancel each other out, and thus the fact that the coexistence of God and evil is logically possible should not influence our estimation of the likelihood that God exists. If we had independent evidence for an all-good, all-powerful, and all-knowing God, we might have to accept God's existence despite any reservations raised by the Darwinian problem of evil. As we've seen in previous chapters, though, no such evidence is forthcoming. Furthermore, even if it were, we would still be stuck with the very real possibility that, if there is a God, he is evil or impotent or ignorant or some combination of these things.

Theism struggles to explain evil and suffering and, when it comes up short, declares it a mystery or a test of faith. In contrast, an atheistic Darwinian worldview provides a compelling explanation for the existence of gratuitous suffering in the world, an explanation that involves fewer contortions of the intellect than any theistic alternative. Furthermore, it puts us in the position where we can finally answer an ancient and vexing question: why do good people suffer? Simple: because the capacity to suffer was crafted by the 'callous' hand of natural selection and sometimes that capacity is activated. Sometimes it's grossly and unfairly over-activated. And that's all there is to it. This is not the kind of answer that those posing the question are searching for, but that is irrelevant to the question of whether it's true or not.

Where are we? Our main conclusion so far is that evolutionary theory takes away some of the major reasons for believing in God, and that the Darwinian version of the problem of evil suggests either that there is no God or, at best, that God is not quite the supreme being we imagined him to be. But perhaps we're not giving theists a fair hearing. Perhaps the conceptions of God we've been attacking up until now don't represent the best they have to offer. This is a common retort, and the subject of the next chapter.

SEVEN

Wrapping up religion

We have thus arrived at the answer to our question, What is Darwinism? It is atheism. This does not mean ... that Mr Darwin himself and all who adopt his views are atheists; but it means that his theory is atheistic; that the exclusion of design from nature is ... tantamount to atheism.

Charles Hodge (1874), pp. 176–7

The true evolutionary epic, retold as poetry, is as intrinsically ennobling as any religious epic. Material reality discovered by science already possesses more content and grandeur than all the religious cosmologies combined. The continuity of the human line has been traced through a period of deep history a thousand times older than that conceived by the western religions.

E. O. Wilson (1998), pp. 289–90

If oxen and horses and lions had hands and could draw as man does, horses would draw the gods shaped as horses and oxen like oxen, each making the bodies of the gods like their own.

Xenophanes

Humanizing nature

Up until now, the conceptions of God we've been dealing with have all had one thing in common: they are traditional, anthropomorphic conceptions of the deity. In other words, they all involve a God that in some sense resembles a human being. Admittedly, these conceptions are not as overtly anthropomorphic as the gods and goddesses of ancient Greece or traditional Hinduism – or, for that matter, the original Abrahamic God, from which today's more rarefied God concepts have descended. Nonetheless, it's fair to say that God, as most people understand him, is an anthropomorphized being. He engages in activities such as creating and designing – activities in which humans also commonly engage (albeit on a more modest scale). He partakes in humanlike social interactions: he communicates with the faithful and listens to their prayers; he returns good for good and bad for bad; and he functions in a wide range of social roles, including leader and lawgiver, father and friend. Finally, he has humanlike mental states: he approves, disapproves, and forgives; he has a loving nature; and he wants to be known and revered (like a teenager craving fame). Of course, God not only has humanlike traits, he has them in superhuman proportions. Thus, whereas we have partial knowledge, God is omniscient; whereas we exert some degree of power over the world, God is omnipotent; and whereas we are partially good, God is as good as it gets. One might say, then, that God is an idealized superhuman mind. But the key point remains: the traditional God may long ago have shed his body, but he is still anthropomorphic to the core.[1]

The arguments we've considered in past chapters apply largely to the traditional conception of God. It is always open to theists to object that this is a dated, childish, and simplistic conception, and

---

[1] The anthropologist Stewart Guthrie (1994) makes a strong case that religious beliefs actually have their origin in an evolved human tendency to anthropomorphize: they are accidental by-products of this tendency. For other attempted evolutionary explanations of religious belief, see Atran (2002); Barrett (2004); Barrett and Keil (1996); Bering (2005); Boyer (2001); Darwin (1871); Dennett (2006); Hamer (2004); Kelemen (2004); Kirkpatrick (2005); D. S. Wilson (2002).

one which fails to describe the God in whom they believe. Many do take this position. They accuse sceptics of promulgating a caricature of God, and chide them for attacking outmoded conceptions that don't represent the real (or at least the best) views of their religion. The anthropomorphic conceptions are either distorted visions of God, or at best metaphors designed to put the true reality of God partially within the reach of the unlettered masses. It's only fair to point out that most believers *do* hold anthropomorphic conceptions, and thus that to attack such conceptions is to attack religion as it really is, rather than as an intellectual minority thinks it ought to be. Nonetheless, it is also only fair to examine some of the non-anthropomorphic conceptions and to assess the claim that, properly understood, God is not susceptible to the kinds of challenges from evolutionary theory that we've been exploring so far. That's the task of this chapter, which is the final chapter in this section.

Renovating the Almighty

If the traditional God is not the true God, what exactly is? What do sophisticated theists really believe? The answer – which itself does little to enhance the credibility of this line of thinking – is that they believe a very wide range of different things. We've already come across one non-anthropomorphic conception: the God of the deists, if stripped of the usual baggage, is strictly a First Cause or Unmoved Mover rather than anything like a person. Another non-anthropomorphic conception involves identifying God with nature. This is known as *pantheism*. For pantheists, the universe is not the creation of God; the universe *is* God. Alternatively, the universe is the body of God and God is the Soul of the universe, like a ghost in a machine. Some of the early religious responses to evolutionary theory leaned heavily towards pantheism (or its close cousin, *panentheism*). For instance, in a description of the impact of evolutionary theory on the ongoing dialogue between science and religion, the Anglican clergyman Aubrey Moore wrote:

> Science had pushed the deist's God farther and farther away, and at the moment when it seemed as if He would be thrust out altogether, Darwinism appeared, and, under the disguise of a foe, did the work of a friend. It has conferred upon philosophy and religion an inestimable benefit, by showing us that we must choose between two alternatives. Either God is everywhere present in nature, or He is nowhere.[2]

Needless to say, Moore's conclusion was that God is everywhere. In his view, we shouldn't seek to find God in the gaps left by science or in miraculous deviations away from natural law. God is to be located in the orderly flow of nature – in every atom, every event, every regularity unearthed by scientific investigation. Put simply, God is immanent in his creation. And as Henry Drummond, a nineteenth-century evangelist, suggested, 'an immanent God, which is the God of Evolution, is infinitely grander than the occasional wonder-worker, who is the God of an old theology'.[3]

Another flavour of non-anthropomorphic deity was promoted by theologians such as Paul Tillich and Hans Küng. The ideas of Tillich have become particularly popular. God, he claimed, is not one being among many but is 'being itself' or the 'ultimate ground of being'. Tillich maintained that *it is false to say that God exists*, but that it is also false to say that he does *not* exist. This is because God is *beyond* existence and non-existence; he is the foundation or sustainer of all that exists. If this seems a bit vague, maybe Küng can help clear things up. According to Küng, God is '*the absolute-relative, here-hereafter, transcendent immanent, all-embracing and all-permeating most real reality in the heart of things, in man, in the history of mankind, in the world*'.[4] Like Tillich's God, Küng's transcends the binary categories of human

---

[2] Moore (1891), p. 73.
[3] Drummond (1894), p. 334.
[4] Küng (1980), p. 185.

thought and language. Thus, if asked whether God is a personal being or an impersonal something, Küng would respond that he is neither. God transcends categories such as personal and impersonal. He is *supra*personal.

More recently, the theologian John Haught proposed a conception of God not as creator or designer but as 'open future'. This conception, he argued, is perfectly consistent with evolutionary biology.

> A God whose very essence is to be the world's open future is not a planner or a designer but an infinitely liberating source of new possibilities and new life. It seems to me that neo-Darwinian biology can live and thrive quite comfortably within the horizon of such a vision of ultimate reality.[5]

Others have attempted a more radical redefinition of God. Following in Wittgenstein's hallowed footsteps, for example, some argue that God is not a propositional belief – a statement to be judged true or false – but is instead an orientation to life or a commitment to a particular lifestyle or group.[6] (I suspect most believers would be very surprised to learn that God is not a propositional belief!) Finally, there are those who claim that their God is a non-anthropomorphic one, but who define God in only the vaguest of ways, if at all. Some view God as no more than the explanation for the strange fact that there is something rather than nothing, and accept that whatever this is, it is unlikely to be anything like a person. Some profess merely to believe in 'something': some kind of higher power or higher purpose. And some maintain that the true nature of God is simply beyond human ken. The seventeenth-century philosopher Blaise Pascal, for example, wrote that: 'If there is a God, He is infinitely incomprehensible, since, having neither parts nor limits, he has no affinity to us. We are then incapable of knowing either what He is or whether

---

[5] Haught (2000), p. 120.
[6] See, e.g., Barbour (1997).

He is.'[7] In other words, God is transcendent, and we can say nothing about his true nature.

That's just a sampling of the non-anthropomorphic conceptions on offer. All have two things in common. First, they involve the rejection of the view that God is a person or person-like being, a supernatural creator akin in some sense to a human mind. And, second, they offer one last chance for a reconciliation between God and evolutionary theory. Let's see how they fare.

Implications of evolutionary theory

The non-anthropomorphic Gods are immune to most of the arguments we considered in earlier chapters. The problem of evil, for example, poses no threat to a non-anthropomorphic God, for it is only if God is an omnipotent moral agent that the existence of evil is problematic. Thus, the main implication for the religious believer here is that, just as the fact of evolution urges a non-literal interpretation of the biblical stories, the Darwinian problem of evil urges a non-traditional conception of God (or, of course, the rejection of God altogether – let's not forget that option).

This is, however, an extremely important implication, for the non-traditional conceptions come at a very high cost. First, there are major problems with the conceptions themselves. Unlike traditional views of God, it is virtually impossible to get a handle on the non-anthropomorphic ones. The ideas of Tillich and Küng, in particular, are so vague and abstract that it's not clear whether they really mean anything at all. One could be forgiven for thinking that they're nothing but gibberish. Thus, evolutionary theory forces us to choose between atheism and gibberish. To be less blunt about this, the theory pushes us either to abandon belief in God or to adopt a conception of the divine so vague that it is virtually exempt from rational discussion

---

[7] Pascal (1910), p. 84.

or debate. And that's really the whole point of these conceptions. They are unfalsifiable hypotheses by design; they've been made unfalsifiable to protect them from sceptical attack. As we've seen, some believers are quite open about the fact that their conceptions of God are vague. Pascal, for instance, claimed that we are utterly incapable of knowing what God is. But this is a very unfortunate position to have to adopt. Anyone who does adopt it is, in effect, claiming that they believe in *something*, but they don't know what. God becomes an empty placeholder, a conceptual black box devoid of any content. We are meant to love God but apparently God is unknowable. If we love someone we don't really know, isn't that just infatuation?

Even if we turn a blind eye to these problems, it is not clear that the non-anthropomorphic conceptions do the sort of work people want and expect from God. Why would anyone pray to the laws of nature, or the Ground of All Being, or the most real reality in the heart of things? Why would anyone feel that any of these things are deserving of worship? A God stripped of all humanlike traits is not a God with whom one could have a personal relationship. It is not a God that could be responsive to our prayers or wishes, or that could love or care for us. It is difficult to see that such a God could guarantee that human standards of justice will prevail in the universe, or that goodness will eventually prevail over evil. A non-anthropomorphic God is so distant and obtuse that it's unable to explain anything (if you're still interested in that). If God is defined simply as the Ground of All Being or the ultimate explanation for existence, we would have no reason to think that God's existence implies that human life is privileged in the universe or ultimately meaningful, or that the universe is greater or more sacred because God exists than it would be otherwise. It seems you could take little comfort from such a conception of God. And if you *do* take comfort from this conception, you should ask yourself whether you genuinely do believe in a non-anthropomorphic God, as opposed to a more traditional one.

On top of all this, the non-traditional conceptions of God are so far removed from the original meaning of the term that it is highly questionable whether it is appropriate to use that term any longer. The singer Frank Sinatra once said: 'I believe in nature, in the birds, the sea, the sky, in everything I can see or that there is real evidence for. If these things are what you mean by God, then I believe in God.' And so does every sane person. But you might just as well say 'I believe in Swiss cheese; if *that's* what you mean by God, then I believe in God.' The question becomes: is it *appropriate* to define God as Swiss cheese – or nature, or the lawfulness of the universe, or the totality of all that exists, or the ultimate explanation for things? And a useful rule of thumb is that if an atheist would not dispute the existence of the thing in question (as is the case with all these examples), it is not appropriate to give that thing the name 'God'. Indeed, the case could be made that those who hold non-anthropomorphic conceptions have long since let go of God, but just haven't admitted it yet – maybe not to themselves, certainly not to others. In the nineteenth century, Ludwig Feuerbach described the kind of philosophies we've been discussing as 'a subtle, disguised atheism', and he had a point.[8] Here's one of my favourite quotations on the topic of religion; it comes from Sigmund Freud:

> In reality these are only attempts at pretending to oneself or to other people that one is still firmly attached to religion, when one has long since cut oneself loose from it. Where questions of religion are concerned, people are guilty of every possible sort of dishonesty and intellectual misdemeanor. Philosophers stretch the meaning of words until they retain scarcely anything of their original sense; by calling 'God' some vague abstraction which they have created for themselves, they pose as deists, as believers, before the world; they may even pride themselves on having attained a higher and purer idea of God,

---

[8] Dowd (2007), Goodenough (1998), and Kaufman (2008) all strike me as examples of this phenomenon, although I'm sure they'd disagree.

although their God is nothing but an insubstantial shadow and no longer the mighty personality of religious doctrine.[9]

Much of what Freud said must be taken with a grain of salt; in the above quotation, however, the old curmudgeon really hits the nail on the head. A certain type of person is happy not to believe in God, but will do anything to avoid having to say 'I don't believe in God.' There comes a point, though, when you should stop trying to redefine the word, and simply concede that you're an atheist. At the very least, you should concede that you're an atheist with respect to the God of the vast majority of believers. Alternatively, if you insist on (mis)using the word God, you should concede that the vast majority of religious people *don't* believe in God[10] – after all, they certainly don't believe in anything like the God you claim to believe in. And nor, in all probability, did the authors of the world's sacred texts or the prophets and priests of the ages, whose God was clearly an anthropomorphic one. Bertrand Russell once pointed out that 'People are more unwilling to give up the *word* "God" than to give up the idea for which the word has hitherto stood.'[11] Evolutionary theory may not persuade everyone to give up the word. However, to the extent that it encourages people to alter its meaning beyond recognition, it could be argued that God has nonetheless been a casualty of Darwin's theory.

Before putting God aside and moving on to our next topic, let's take a moment to take stock. In the wake of Darwin's theory, there are five major positions available on the question of God and evolution. (1) You could maintain a belief in a traditional God and reject both the fact and the theory of evolution (the Creationist position). But

---

[9] Freud (1927), pp. 28–9.
[10] This was actually Tillich's view. In his opinion, the God of the multitudes was pure superstition. Not surprisingly, various commentators argue that Tillich was really an atheist (e.g., Hook, 1961).
[11] Russell (1935), p. 215.

the evidence for evolution is so strong now that this position is as intellectually respectable as the view that the earth is flat. (2) You could maintain a belief in God and assume that God personally guided the process of evolution. This involves acceptance of the fact of evolution, but is inconsistent with the *theory* of evolution, and leaves us scratching our heads about the imperfections we find in the design of living things. (3) You could maintain a belief in God, and assume that God chose natural selection as his means of creating life by proxy. This involves accepting both the fact of evolution and the modern theory of evolution, but it also involves abandoning the God of the monotheistic religions – something most believers will be unwilling to do. Furthermore, you must face the problem of evil, and evolutionary theory adds fuel to the fire of this ancient dilemma. (4) You could maintain a belief in God and accept evolutionary theory, but replace a traditional conception of God with a non-anthropomorphic one. This avoids the problem of evil. However, it reduces God to a distant and impersonal abstraction, and raises the question of whether it makes any sense to call this abstraction *God*. (5) Last but not least, you could accept both the fact and the theory of evolution, and reject belief in God in any way, shape or form. After Darwin, this is the least problematic option.

PART II

Life after Darwin

EIGHT

Human beings and their place in the universe

The fact that we are the contingent end-products of a natural process of evolution, rather than the special creation of a good God, in His own image, has to be just about the most profound thing we humans have discovered about ourselves.

Michael Ruse (1986), p. xi

Let us now consider man in the free spirit of natural history, as though we were zoologists from another planet completing a catalog of social species on Earth. In this macroscopic view the humanities and social sciences shrink to specialized branches of biology ... anthropology and sociology together constitute the sociobiology of a single primate species.

E. O. Wilson (1975), p. 547

Man has been here 32,000 years. That it took a hundred million years to prepare the world for him is proof that that is what it was done for. I suppose it is. I dunno. If the Eiffel tower were now representing the world's age, the skin of paint on the pinnacle-knob at its summit would represent man's share of that age; & anybody would perceive that that skin was what the tower was built for. I reckon they would. I dunno.

Mark Twain, 1903, cited in Twain (1992), p. 576

Breaking down the walls

Science has turned our view of ourselves and the world we live in upside down and inside out. People living today – even those with little formal education – have a radically different view of the nature of the universe than their counterparts in earlier centuries. This is entirely the result of the scientific enterprise. The classic example of science overturning our natural view of things is the discovery that the earth orbits the sun, rather than the other way around. This undermined what was once thought to be a fundamental dichotomy in nature: the distinction between the earth and the heavens – the world in which we live and the inaccessible realm we glimpse above us on cloudless nights. To our forebears, this distinction seemed utterly obvious and utterly unassailable. But once Copernicus and Galileo showed us that the earth is in orbit around the sun, it crumbled like dust. The earth is not separate from the heavens but is *in* the 'heavens'; it is one celestial object among many, occupying a non-special position in the interstellar darkness. The earth–heaven dichotomy disintegrated, and people had to make a major readjustment to their construal of the world.

Given the current state of our knowledge, it is not possible to fully grasp what a radical shift in perspective this was. The same cannot be said for Darwin's theory of evolution. According to modern evolutionary theory, human beings and all other species are products of a mindless process of gene selection, products of the differential survival of lengths of DNA. This represents a dramatic departure from our previous convictions about our origins, and not surprisingly the implications don't end there. In this and the next few chapters, we'll survey the implications of evolutionary theory for our understanding of who and what we are. We'll look at the place of human beings in the universe, our status among the animals, and the meaning and purpose of human life. First of all, though, in this chapter, we'll look at how evolutionary theory challenges pre-Darwinian ideas about ourselves and our relationship to the natural world. We'll see that, just as the Copernican revolution broke down the wall between the earth and the heavens, the Darwinian revolution threatens to break down the walls between mind and matter, humans and animals, life and non-life. And we'll see that, in many ways, these implications are more unsettling and disturbing than those of the work of Copernicus.

Life before Darwin

To begin with, I'll sketch a rough outline of some important pre-Darwinian trends of thought. One recurring idea in the belief systems of the world is the notion that human beings are composed of two separate and separable parts: a physical body and a mind (or soul). Mind–body dualism is not a view that all peoples have held, but it is also not a uniquely Western view, as some have claimed. It is implicit in such beliefs as that the spirit of a tribal shaman can venture beyond the body and visit spiritual realms; that a person's body can be taken over and controlled by another mind; and that the same individual can be born into a series of different bodies. The mind–body distinction took on a particular importance in traditional Western theology, according to which the mind is a spark of divinity, a part of us that transcends the material world and our biological natures. Because the mind is something separate from nature, humans are not just physical entities. We combine heavenly and earthly aspects, immaterial souls and material bodies; we have one foot in the world of matter and one in the world of spirit.

The two-part model of our makeup is closely associated with another persistent belief among members of our species: that in one way or another we survive physical death. This belief comes in many different shapes and sizes. To begin with, a distinction can be drawn between survival beliefs that posit continued existence outside the body, and those that posit continued existence within the body. Survival outside the body is variously conceived as survival in an astral or ghost body, or survival as a disembodied mind. A popular belief along these lines is that, at death, the soul disentangles itself from the body and migrates to an after-world (e.g., heaven, hell, the happy hunting ground). There are also a number of ideas about survival *within* a physical body. One is the traditional Judaeo-Christian and Islamic doctrine that God will resurrect our bodies in the future, at which time we will face his judgment for our conduct in this life. Another is the doctrine of reincarnation, which is found among Hindus, Buddhists, and many New Agers. All these conceptions of life after death have in common the fact that the individual person survives in some sense. This is not a feature of all survival beliefs, however. Some strains of Buddhism, for example, hold that the individual mind ultimately merges back into a universal mind – that in death, we sink back into the state of inorganic matter and are reabsorbed into the oneness of everything.

As well as dividing the world into mind and matter, before Darwin it seemed meaningful to divide the inhabitants of this planet into humans and animals. Humans were held to have a radically different nature from animals. According to Western theology, people have immortal souls, whereas animals do not. As long as we're well behaved, or hold the correct beliefs about God and his offspring, we can live forever in heaven after we die. This option is not available to animals. (Family pets are an apparent exception, especially those survived by heartbroken children.) Not only are we alone immortal, we alone are made in God's image and have a special relationship with the deity, and we alone know God and moral truths. In an address on evolution to the Pontifical Academy of Sciences in 1996, Pope John Paul II, stated that 'man is the only creature on earth that God wanted for its own sake'. (He didn't explain how he came by this knowledge.) In a slightly different vein, many ancient Greek, Christian, and Enlightenment thinkers held that humans are distinguished from animals by the spark of reason. We can talk, count, and reason deductively; animals cannot. Some thinkers even claimed that humans alone are conscious. Descartes taught that animals are non-conscious automata, incapable of thought and devoid of the experience of pleasure or pain.[1] A less extreme and more widely held view is that animals are conscious but not *self*-conscious – they know but they don't know that they know. Regardless of the details, though, all these views posit a fundamental distinction between human beings and animals.

A subtler aspect of the pre-Darwinian worldview concerns the reality of species. Before Darwin, philosophers believed that species were ontologically fundamental divisions in nature, as unchanging and immutable as numbers or geometrical shapes.[2] Our species concepts 'carve nature at the joints', to use Plato's memorable phrase, and in that sense species are real. This position is consistent with the common-sense view on the topic. Most people assume that exact boundaries can be drawn around each species category, such that everything in the universe can be placed either within or outside the category. So, just as it's possible to classify every object in the universe as a triangle or not a triangle, it's possible to classify every object as a member of a given species or not a member: a human or not a human; a cat or not a cat; a fir tree or not a fir tree. Similarly, every object can be classified as animal or not animal, living or not living. There are no in-betweens or grey areas. So says common sense.

Another widespread and intuitively plausible component of the pre-Darwinian worldview is the idea that life is distinguished from non-life by the possession of an invisible essence or vital spirit. If you drop a dead bird from the top of a building, it plummets to the ground in a manner that can be explained purely in terms of the laws of physics. But what happens if you drop a live bird from the top of a building? It begins to flap its wings, pauses momentarily to get its bearings, and then flies in a graceful arc to settle in a nearby

[1] Descartes (1986).
[2] Dennett (1995).

tree. No simple formula in physics predicts this sequence of events. Can we really believe that the behaviour of life forms – especially animals, especially human beings – is explicable solely in terms of the action of physical forces on matter? It seems that another force must be at work: a life force, or what the philosopher Henri Bergson called the *élan vital*.[3] This vital force acts on inert matter, breathing life into the lifeless molecules of the universe, inspiring them to dance up in defiance of the laws of physics and their simplistic prescriptions for matter. Advocates of this position are known as *vitalists*. Vitalists deny that humans or animals are merely complex physical machines. Instead, they are animated by a soul or life force, and it is this that distinguishes them from the non-living portions of the universe.

Giving up the ghost

The above ideas touch on many different areas of thought. Evolutionary theory, on the other hand, has only one goal: to provide a naturalistic account of the origin of the varied life on this planet. In this respect, it clearly contradicts all pre-Darwinian views about how we came to exist. Most dramatically, it undermines the creationist view that we were created in our present state in the geologically recent past. Rather than Adam and Eve, our ancestor was, in Darwin's words, 'a hairy quadruped, furnished with a tail and pointed ears, probably arboreal in habits, and an inhabitant of the Old World'.[4] This creature itself evolved from other mammals, which evolved from reptiles, which evolved from amphibians, which evolved from fish. Ultimately, all of us can be traced back to inanimate chemicals. Evolutionary theory is inconsistent with the traditional Abrahamic creation myth, and indeed with the creation myths of all cultures, in as much as these are taken literally.

---

[3] Bergson (1911).
[4] Darwin (1871), p. 678.

This much is obvious. But what else follows from an evolutionary account of human origins? Well, for a start, evolutionary theory threatens to break down the wall between mind and matter. As we've seen, the theory doesn't just provide a naturalistic explanation for physical structures, such as eyes and wings and teeth and claws; it provides a compelling account of the origin of the mind. According to evolutionary psychologists, the mind, like any other complex aspect of the organism, is the product of a slow and gradual accumulation of variants that, at each stage in the process, enhanced the fitness of the individuals possessing them. Different components of the mind evolved because they contributed in different ways to the survival and reproductive success of mind-endowed organisms. For example, just as lions evolved large teeth because they helped them to tear apart their prey, and just as gazelles evolved fast legs because they helped them to outrun lions, the emotion of fear evolved because it motivated our ancestors to avoid or escape threats to life and limb; sexual desire evolved because it led our ancestors to engage in behaviours resulting in the production of babies; and parental love evolved because it motivated our ancestors to care for these babies once they arrived on the scene. In short, the mind is a device designed to perpetuate the genes contributing to its development – or, to look at this from another angle, minds are devices that our genes have thrown into existence in order to perpetuate themselves.

Evolutionary psychologists have made a strong case for this view.[5] They've pointed out, among other things, that the mind possesses certain propensities and quirks that make sense only on the assumption that it was crafted by natural selection.[6] For instance, many of our fears are tuned to specific threats that were relevant to our hunter-gatherer ancestors, but that are now largely irrelevant. Even people

[5] See, e.g., Barkow et al. (1992); Betzig (1997); Buss (1995); Daly and Wilson (1988); Dunbar and Barrett (2007); Kenrick and Luce (2004); Miller (2000); Pinker (1997); Ridley (1994); Trivers (2002); Wright (1994).
[6] Marcus (2008).

living in industrialized conditions are more likely to develop phobias about ancestral threats, such as snakes and spiders, than they are to develop phobias about items much more dangerous to them in the modern environment, such as electrical sockets, fast driving, or hair dryers near bathtubs. Our fear of snakes and spiders is an example of an aspect of human psychology that is poorly matched to modern living conditions, but which would have been useful in the environment of our hunter-gatherer ancestors – the environment in which these fears evolved.[7] If the mind were a product of divine design, why would God only instil in us an instinctive fear of ancestral threats? Natural selection has an excuse: it does not have foresight and sometimes the environment changes so fast that selection cannot keep up. But God has no excuse, because God is supposedly infinitely foresightful. This doesn't rule out the existence of God, but it does tell us that, if there is a God, he left the task of creating and designing minds up to natural selection. God or no God, the mind evolved.

This has a number of implications. Several of these stem from the fact that, from an evolutionary perspective, the mind is the activity of the brain.[8] Given that the brain is an entirely mechanical contraption, it would seem that the mind must be just as mechanical as the brain. This would appear to rule out free will, even if free will were possible in principle.[9] A more disturbing implication is that the mind could not survive the death of the body. Darwin once wrote: 'As for a

---

[7] In saying this, I am not endorsing any particular version of the 'Stone Age Mind thesis' – the idea that our minds are 'designed' primarily for the hunter-gatherer environment of our Pleistocene ancestors. I do think that it's possible to frame a sensible version of this thesis; for present purposes, though, all I am claiming is that *some aspects* of our psychology evolved in response to selection pressures that no longer exist, and thus that they are not well matched to our modern environment. See Tooby and Cosmides (1992) and Sterelny (2003) for discussions of this issue.

[8] See Clark (2008), Fish (2009), and Noë (2009) for an alternative naturalistic view of the mind, according to which the mind is best understood as an environment–organism relationship, rather than a species of brain activity.

[9] Although see Dennett (2003).

future life, every man must judge for himself between conflicting vague probabilities."[10] But the case against survival is stronger than Darwin here allows. All the evidence points to the fact that the mind is dependent on the activity of the evolved brain. Neuroscientists have shown that when you look at something – when you have a conscious visual experience – certain parts of your brain become more active. If you then close your eyes and merely *imagine* the same visual scene, the same parts of your brain again become active. This shows that not only does perception have an identifiable physical basis, so too does imagination. If you electrically stimulate the visual areas of the brain, this produces conscious visual experiences. Stimulating other sensory areas produces other sensory experiences. Other things that influence brain states, such as recreational drugs, simultaneously influence states of mind. It seems that everything we are conscious of – every sensation, feeling, recollection, or thought – is associated with activity in the brain (or rather *is* activity in the brain). Furthermore, as David Hume noted several centuries ago in his essay, *Of the Immortality of the Soul*:

> The weakness of the body and that of the mind in infancy are exactly proportioned; their vigour in manhood, their sympathetic disorder in sickness, their common gradual decay in old age. The step further seems unavoidable; their common dissolution in death.[11]

We know that when part of the brain is destroyed, so too is part of the mind. Can we believe that when the brain is *completely* destroyed, the mind, rather than being completely destroyed also, is instead completely restored? Without a strong reason to think that this is the case, it is much more reasonable to assume that our conscious existence ends with the death of the brain. The fact that the mind is dependent on the brain essentially rules out survival outside a physical body. It also rules out reincarnation, as reincarnation requires the persistence

[10] Cited in F. Darwin (1887a), p. 277.
[11] Hume (1777), III.X.33.

of a mind without a brain between incarnations – a mind that can be transferred from one brain to another. Furthermore, as the philosopher Robert Nozick pointed out, even if survival of death were possible in principle (a doubtful proposition), 'there would be no selective pressure for the survival of bodily death, for this would not lead to greater reproductive success'.[12] It seems that the evolutionist must conclude, along with the writer Vladimir Nabokov, that 'our existence is but a brief crack of light between two eternities of darkness'.[13] Brains that think otherwise – brains that deny they are brains and believe instead that they are eternal souls – are brains that hold false beliefs about themselves.

Having established this point, we might want to ask why people so consistently fear death. If death is nothing, then surely there is nothing to fear. We do not regret our past non-existence, so why do we regret the prospect of our *future* non-existence? We do not mourn the fact that our loved ones did not exist before they were born, so why do we mourn their non-existence after they die? To a purely logical mind, these would seem to be reasonable questions. But given the importance of survival for evolved beings, it is hardly surprising that we instinctively fear death and mourn our loved ones. According to the philosopher Derek Parfit, 'In giving us this attitude, Evolution denies us the best attitude to death.'[14] The fear of death is an unpleasant – and ultimately unfounded – gift from natural selection.

So one implication of the evolutionary psychologist's view of the mind is that it rules out life after death. A less obvious implication concerns the dividing line between mind and matter. We naturally think that everything in the universe either has a mind or does not, is conscious or is not. But the fact that the mind evolved suggests that there is no clear and non-arbitrary line to be drawn between mind and

---

[12] Nozick (2001), p. 360.
[13] Nabokov (1967), p. 17.
[14] Parfit (1984), p. 177.

non-mind.[15] The evolution of minds must have been a slow and gradual process, like the evolution of species. It must have involved a continuum of forms spanning from the mindless to the fully minded, with many intermediate stages that could not be classed one way or the other, not because we don't know enough to decide whether the intermediate stages count as minds or not, but because there is no non-arbitrary answer to this question. (Note, incidentally, that we should be cautious of the idea of 'full mindedness'. To a hypothetical alien with an intellect vastly superior to our own, human minds would be classed as intermediate forms between the mindless and the fully minded.) Just as the Copernican revolution undermines the distinction between the earth and the heavens, the Darwinian revolution undermines the distinction between mind and matter. Mind is not something distinct from matter; mind *is* matter, arranged and functioning in a particular way. The seat of reason is not an immaterial something but a mechanism: the brain.

Some people find this view demeaning. 'How can you feel good about yourself if you think you're just a computer made of meat?' they ask. But in a certain sense, an evolutionary perspective on the mind makes it *more* amazing, not less. As the neurobiologist Marian Diamond put it, 'The brain is a three pound mass you can hold in your hand that can conceive of a universe a hundred billion light years across.' It's quite an amazing thing when you think of it that way, especially when you consider that this three-pound mass was not assembled by a conscious being but by a mindless natural process.

The conscious universe

We see, then, that mind is not something separate from matter; mind is a process embodied *in* matter. With this new perspective under our

---

[15] Dennett (1991).

belts, it becomes impossible to maintain that the mind stands outside nature. Instead, mind becomes a tiny *fragment* of nature, valued only by those tiny fragments of nature that possess it. When we fully digest this idea, it radically transforms our view of the mind's place in the universe – and our view of the universe itself. The physical universe ceases to be an unconscious object, observed and explored by conscious minds which somehow stand above or outside it. Conscious minds are *part* of the physical universe. As soon as we recognize this, we realise that the universe itself is partially conscious. When you contemplate the universe, part of the universe becomes conscious of itself. Similarly, our knowledge of the universe is not something separate from the universe; it is a *part* of the universe. Thus, for humans to know the universe is for the universe to know itself. As Carl Sagan put it, 'humans are the stuff of the cosmos examining itself'.[16] And Darwin's theory explains how this could be so – how clumps of matter could come to be organized in such a way that they are able to contemplate themselves and the rest of the cosmos.

The history of the universe looks very different from this perspective. For billions and billions of years, the universe was here and no one knew about it. More to the point, for all that time, the universe itself had no idea that it existed. But then, around 13.7 billion years after the Big Bang, and almost four billion years since life first evolved, something strange began to happen: tiny parts of the universe became conscious, and came to know something about themselves and the universe of which they are a part. (Note that this could have happened elsewhere in the universe as well, and perhaps much earlier.) Eventually, some of these tiny parts of the universe – the parts we call 'scientists' and 'scientifically informed laypeople' – came to understand the Big Bang and the evolutionary process through which they had come to exist. After an eternity of unconsciousness, the universe now had some glimmering awareness that it existed and some

---

[16] Cited in Munz (1993), p. 185.

understanding of where it had come from. This might sound like a strange thing for a universe to do, but perhaps it's not; perhaps many possible universes would become conscious of themselves given sufficient time.

So much for the history of the conscious universe; what about its destiny? There are many competing suggestions on this topic, some more optimistic than realistic. Pierre Teilhard de Chardin, a French Jesuit palaeontologist, suggested that the universe will continue to expand into greater and greater degrees of awareness, finally coalescing into an integrated, universal consciousness, which he dubbed the *Omega Point* and identified with Christ.[17] Modern cosmology indicates that such suggestions are more interesting than they are plausible. Although the universe is conscious of itself at present, the projected heat death of the universe makes it all but certain that the time will come when the lights will go out and the universe will slip back into unconsciousness. For how long will it remain in its present semi-conscious state? The answer depends on how prolific the universe is at producing conscious life. If consciousness is widespread throughout the universe, then the odds are that at least some pockets of consciousness on some planets will survive for a reasonable length of time. For all we know, though, ours may be the only planet in the universe hosting mind and consciousness. If so, then our decisions and our conduct will determine whether the universe has a long future as a conscious entity or will soon lapse back into unconsciousness.

That said, one might wonder whether, in the grand scheme of things, it really matters. It may be pure anthropocentrism to assume that a universe with consciousness is better than one without. Conscious beings are often disgruntled and sometimes simply miserable, and maybe on balance an unconscious universe would be the more desirable. But although it's possible to entertain such thoughts in principle, it's hard in practice to duck the conclusion that it would

[17] Teilhard de Chardin (1959); see also Tipler (1994).

be a terrible shame if the universe were not to remain conscious for as long as possible. Nonetheless, it may be the fate of the universe to spend an eternity in darkness, save one brief flash of self-awareness in the middle of nowhere.

Humans and animals

Our main conclusion thus far is that evolutionary theory challenges the distinction between mind and matter and that, in the process, it recasts us as entirely natural creatures. This, in turn, urges a thorough rethink of our relationship to other animals. Not only does Darwin's theory place us squarely within the natural world, it stresses our kinship with the animals. They and we are not products of separate acts of creation. Instead, we are distant relatives. Every human being is a distant relative: your spouse or lover, the stranger who sat next to you on the bus this morning, Einstein, Hitler, Buddha. Chimpanzees and other apes are more distant relatives still. When you go home and stroke your cat or take your dog for a walk, you are interacting with even more distant relatives. When you walk across a grassy field on a sunny day, the grass beneath your feet and the flowers you stop to admire are yet more distant relatives. Ultimately, all known life on this planet can be traced back to a common ancestor, and every life form that has ever existed here can be placed on a single family tree. All life on earth is literally one large family (albeit a dysfunctional family in which the family members have a nasty habit of eating one another). This is one of the most profound lessons of evolutionary theory.

Darwin's theory does more than simply stress our kinship with the animals. It challenges the very idea that the inhabitants of this planet can be meaningfully divided into humans and animals in the first place. We all came about through the same process, and our common origin suggests that we will have more in common with other animals than we previously imagined. Certainly, the human–animal distinction is still workable; after all, we rarely make errors in assigning entities to one category or the other.[18] But after Darwin, the distinction suddenly seems arbitrary – as arbitrary as the equally workable distinction between, say, horses and non-horses. Post-Darwin, it is more natural to think of *humans* as a subset of the category *animal*. The idea that humans are not animals makes precisely as much sense as the idea that the earth is not a planet, or the sun not a star. We can say these things if we want to. However, if we wish to frame an objective view of the universe (i.e., a view that would be equally valid from the perspective of any species on any planet), we must view the sun as a star, the earth as a planet – and humans as animals.

This suggestion is no doubt far less shocking to modern ears than it was in Darwin's day. Nonetheless, it is not clear that most people have fully taken on board its implications. If they had, then perhaps academic disciplines such as sociology and anthropology would be viewed as specialist branches of zoology; medical doctors would be viewed as a subtype of veterinarians (one that specializes in tending to the health needs of just one species); human rights would be viewed as a subset of animal rights; and the socialization of children would be viewed as one example of the training or domestication of animals (making parents and teachers a subtype of animal trainers). These examples aren't particularly serious. They do make a serious point, though, which is that, at least to some extent, we still view ourselves as set apart from the rest of the animal kingdom. For this reason, we may have a moral blind spot when it comes to our 'fellow brethren' (as Darwin once described non-human animals). More on this in Chapter 13.

As we've seen, a lot of people reacted very negatively to Darwin's theory. This was partly because the theory removed the need for God. But that couldn't be the whole story. After all, people could still believe

[18] Rarely but not never. When Europeans first came across the great apes in the sixteenth century, they were uncertain whether or not these creatures should be classed as humans. And when they first started to encounter other races, with radically different cultures from their own, they were uncertain whether or not *these* individuals should be classed as humans.

that God created life *through* the process of evolution (at least if they didn't think too hard about it). The adverse reaction may have been driven in large measure by the fact that a lot of people were insulted by the idea that we are animals. Many felt that Darwin had literally made monkeys out of them, and in a sense he had. But if you think about it, their outrage at this is rather insulting to monkeys! To some degree, the public outcry following the unveiling of Darwin's theory of evolution revealed a deep-seated prejudice against non-human animals. A Ku Klux Klan member would be mortified to learn that he was actually a black man. Many people's reaction to learning that they are actually animals, or actually apes, is the same. (For some reason, they're not so perturbed about being classed as mammals or as living things. Are we thinking straight about this issue?) Another reason people may have been resistant to the idea that humans are animals is that it raises a disturbing possibility: if other animals are no more than complex organic machines, as Descartes and others suggested, then perhaps neither are we. Although a lot of people today are quite at ease with the idea that we are animals, in Darwin's day it was profoundly unwelcome news. And the fact of the matter is that for many people it still is.

Are species real?

We've seen, then, that in addition to challenging the mind–matter dichotomy, evolutionary theory helps to break down the wall between humans and other animals. In fact, it does more than this; it challenges the ultimate validity of our species divisions. So here's our next question: does the word 'species' denote a real category in nature? Or is it a human invention, something we artificially impose on the messy ebb and flow of nature?[19] The answer to this question depends entirely on how we choose to define species, and what we

---

[19] See Dupré (1992), Pigliucci and Kaplan (2006), Sober (2000), and Sterelny and Griffiths (1999) for introductory treatments of this issue. See Sober (2006) and Stamos (2004) for more in-depth discussions.

decide it would mean to say that species are real. There are various approaches we could take. We could say, for instance, that species are real if and only if they are manifestations of unchanging Platonic forms or ideas in the mind of God. Or, if we have a more naturalistic bent, we could adopt the position of Darwin's friend and mentor, the geologist Charles Lyell, who suggested that species are real if and only if they have permanent existence in nature and 'fixed limits, beyond which the descendants from common parents can never deviate from a certain type'.[20]

Given either of these definitions, evolutionary theory would immediately imply that species are not real. The Linnaean classification system is in no sense fixed; it represents a mere snapshot of an ongoing process of change and transformation. Our impression that there is something permanent about species is merely an artefact of our myopic perspective on the history of the universe. We occupy such a tiny sliver of time, relative to the timescale of evolutionary change, that the fluidity and ultimate impermanence of species is all but concealed from us. Furthermore, evolution involves an element of randomness, and thus existing life forms are merely historical accidents. As such, they could not be ontologically fundamental divisions within nature in the same way that, say, electrons or quarks are.

Another way to approach the issue would be to stipulate that species are real just as long as we can place clear borders around them. It is sometimes difficult to put a dividing line between existent species; it can often be done, though, especially with the species we're most familiar with: people, pandas, lions, etc. But again, this is only because we occupy such an insignificant slice of time. The task would be virtually impossible if all life forms that had ever existed were spread out in front of us at once. As Darwin pointed out, rather than sets of clearly defined species, we would find a continuum of forms, and would have no way of saying where one species ended and the

---

[20] Lyell (1835), p. 325.

next began. Take human beings. At one point in evolutionary history, there were animals that were clearly not human; at another point, there were animals that clearly were. In between, however, there were animals that could not be classed as human *or* as non-human. This is not because we don't have the requisite expertise to determine whether they were human or not. It is because the human v. non-human concept is dichotomous but the reality is continuous. This applies to all species concepts. There are no clear dividing lines between species in evolutionary history, any more than there is a clear dividing line between childhood and adulthood, or between spring and summer. In this sense as well, species are not real.

Now this is not to suggest that our species designations are entirely arbitrary. They correspond to relatively discrete sets of recurring patterns in nature at particular moments in time. This is demonstrated by the fact that there is great consistency between cultures in the ways that people divide animals and plants into groups. There is a grain of truth in our species concepts that should be preserved, even if these concepts are only clumsy approximations and apply only during a very narrow window of time. Nonetheless, species boundaries do not have the definiteness or significance we might formerly have attributed to them. This further demolishes the idea of an absolute wall separating humans from other animals.

Life and non-life

So mind is matter, humans are animals, and species are convenient fictions. But we still haven't finished our survey of the damage done to our traditional view of the world by the universal acid of Darwinism. On top of everything else, evolutionary theory helps to break down the wall between life and non-life. We saw in Chapter 5 that, although most of the time we can categorize objects as living or not living, there is no real dividing line between these categories in evolutionary history. This is not to say that the concepts are worthless or that they don't correspond to reality even approximately. Usually they serve us well. Some clumps of matter are clearly alive (you and me for a start); other clumps are clearly not alive (rocks, mountains, and mud spring to mind). But there are yet other clumps of matter that cannot be classified so easily. In fact, they cannot be classified at all. The distinction between life and non-life does not encompass everything there is. The inventory of objects in the universe includes items for which the distinction breaks down and can only be maintained if we assign the items in question to one or the other category arbitrarily. Viruses are the prototypical example. There is no answer waiting to be discovered concerning whether viruses are living or non-living. They are what they are, and they reveal the imperfections (at the very outskirts) of our otherwise useful concepts. The crisp distinction between life and non-life doesn't quite map onto the fuzzy, messy reality.

As well as undermining the life v. non-life distinction, evolutionary theory undermines the notion that there is any kind of life force – any force external to matter that enters into and animates the molecules of the universe. The notion of a life force plausibly originated when our forebears began to ask themselves: what is the difference between a dead person and a living person? Physically, they appear identical; they have the same weight, size, etc. The natural inference, then, is that there must be some non-physical ingredient that makes the difference: some kind of animating spirit that leaves the body at death. This argument has some intuitive pull, even today. So it must have seemed irresistibly compelling to our predecessors, lacking as they did any detailed knowledge of how the body works. But compare the above argument to this one. What is the difference between a functioning toaster and a broken one? Physically, they appear identical; they have the same weight, size, etc. Thus the difference must be that the working toaster has some kind of non-physical animating spirit that leaves the toaster when it breaks down. Applied to toasters and other simple machines, we see how weak the argument is. It only seems stronger when applied to people and other animals because animals are so much more complex than toasters, and because we so desperately want to believe that we and our loved ones survive death.

The machine analogy is apt. In fact, technically it's not even an analogy. The vitalists denied that we are 'mere' machines, but they were wrong. We *are* machines: biological machines designed by natural selection. Certainly, we're very complex machines (complex, at least, relative to the powers of our puny minds). And certainly we are conscious, decision-making machines. But we're machines nonetheless. Science now has a good understanding of the basic mechanism of the human body. There is no longer any need to posit an animating force over and above this mechanism. What differentiates inanimate objects from animate ones is not some ethereal life force. It is the way the parts are organized. Each part is itself non-living, but through the organization of these non-living parts, a living being comes into existence. Life is not dead matter plus some added ingredient. Life is just dead matter organized in a particularly interesting way.

Some people worry that to say we are nothing but matter is to deny that we think or feel. It's not. The strange fact is that, when suitably arranged, *matter thinks and feels*. Those chemical elements, which we all found so dull at school, turn out to be capable, when found in a suitable formation and behaving in a suitable manner, of such interesting things as thinking, loving, hating, and dreaming – and thinking about the fact that they can think and love and hate and dream. Given the traditional conception of matter, people had no choice but to say that life forms were more than mere matter – that they were animated by a life force or individual soul. A more parsimonious and plausible alternative is instead to revise the traditional conception of matter. Matter is capable, all on its own, of life and mind. Thus to say that we are nothing but matter is not to say that we are less than we appear to be. It is to say that matter is *more* than we usually think it to be.

Conclusion

The Copernican–Newtonian revolution broke down the artificial distinction between the heavens and the earth. However, it didn't threaten the distinction between mind and matter, humans and animals, or life and non-life. It therefore had much less impact on our view of ourselves and our place in nature than the Darwinian revolution. Evolutionary theory hit us where it really hurts. It is plainly inconsistent with at least four components of the pre-Darwinian worldview. First, it is inconsistent with the idea that mind and matter are distinct domains of existence. The theory implies that the mind is the activity of an evolved brain, which in turn counts out the possibility of life after death – a conclusion that clashes with most of the religious and spiritual belief systems of the world. Second, evolutionary theory is inconsistent with the idea that the inhabitants of this planet can be meaningfully divided into humans and animals. Just as we have accepted that the earth is one among the planets, and the sun one among the stars, we must now accept that our species is one among the animals. Third, the theory is inconsistent with the idea that species are fundamental building blocks of the world in the same way that subatomic particles or atomic elements are. Indeed, if we take in the full span of evolutionary history, we see that there are no clear dividing lines between any species and its predecessor. And, fourth, evolutionary theory is inconsistent with the view that there is any dividing line between life and non-life, or any kind of life force or soul. These are profound changes to the traditional picture of our species and our place in the universe. But we're just getting warmed up. In the next chapter, we ask: are humans superior to other animals?

NINE

The status of human beings among
the animals

God's noblest work? Man. Who found it out? Man.

Mark Twain (1992), p. 943

Man in his arrogance thinks himself a great work worthy of the
interposition of a deity. More humble and I think truer to consider
him created from animals.

Charles Darwin, cited in Barrett et al. (1987), pp. 196–7

Again and again, we encounter sweeping visions, encompassing everything from the primordial dust cloud to the chimpanzee. Then, at the
very threshold of a comprehensive system, traditional pride and prejudice
intervene to secure an exceptional status for one peculiar primate ...
The specific form of the argument varies, but its intent is ever the
same – to separate man from nature.

Stephen Jay Gould (1980), p. 136

King of the earth

If I were to ask you which plant is superior to all other plants – which
is the noblest and most worthy of plants – you'd probably be hard
pressed to think of an answer. In fact, you'd probably think the
question was decidedly odd. If, on the other hand, I were to ask you
which animal is superior to all other animals, the noblest and most

worthy, this would seem like a perfectly reasonable question. And I can guess what your answer would be. It seems an unshakeable conviction of the greater part of humankind that our species is superior to all others. Shakespeare's Hamlet expressed this view in the following famous words:

> What a piece of work is a man! How noble in reason! How infinite in faculty! In form and moving how express and admirable! In action how like an angel! In apprehension how like a god! The beauty of the world! The paragon of animals! (*Hamlet*, Act II, Scene II)

In this chapter, we'll critically examine the claim that human beings are the paragon of animals. Since Darwin, this claim has been closely linked to the idea that evolutionary change is a matter of ongoing and steady advancement, and that humans are leading the charge of evolutionary progress. So, in addition to looking at the issue of our superiority, we'll consider the question of whether evolution is progressive, pressing forever onwards and upwards. Our conclusion will be that, after Darwin, there are no objective grounds to say that human beings are superior among the animals. In other words, we're *not* superior among the animals.

The Great Chain of Being

The assumption of human superiority is ubiquitous among the belief systems of the world, particularly in the West. In the last chapter, we saw various suggestions about differences between humans and other animals: we alone are made in the image of God, we alone possess the spark of reason, we alone are moral creatures. These claims were not neutral descriptions of differences; they were thought to place human beings above all other animals. And it's not only the philosophers and churchmen who have thought that animals are inferior. For one thing, it is implicit in our everyday language. For example, to say that someone 'acted like an animal' is to criticize that person's

behaviour, and animal names are often employed as terms of abuse
(e.g., bitch, snake, rat). Similarly, throughout history, groups that
have dominated or decimated other groups have referred to them as
mere animals, or as sub-human. This is how the Nazis viewed the
Jews, for example, and they were not exceptional in this.

Certainly, not everyone has been quite as negative about other
animals. In many cultures, certain species are deemed sacred, and
Eastern religions such as Buddhism, Hinduism, Jainism, and Daoism
hold that all life is equally valuable (or at least pay lip service to this
idea). Nonetheless, the habit of looking down our noses at other
animals is not limited to Western thought. In traditional Indian
philosophy, it is held that, as a result of bad karma accrued in this
life, people may return as 'lower' animals in their next incarnation;
the classic example is the dung beetle.[1] Implicit in this belief is the
notion that other animals are inferior to us, and thus that it would be
undesirable to be one.

The view that humans are superior among the animals is embodied
in a pervasive idea about the structure of the world, known as the
Great Chain of Being.[2] The Great Chain of Being is based on the
assumption that everything in the universe can be ranked from lowest
to highest. Everything exists on a continuum spanning from pure
matter at the bottom (e.g., rocks and mountains), to pure reason or
spirit at the top (i.e., God). Each higher rung on the ladder represents
a higher 'level of being'. At the bottom is the non-living world, the
world of inanimate objects. Inanimate objects have just one attribute:
existence. Plants occupy the rungs immediately above this. Like
inanimate objects, plants have existence, but they also have life.
Animals occupy the rungs above plants. Like plants, animals have

---

[1] The problem with this idea (aside from the fact that it's obviously not true) is that
so-called lower life forms are often happier and less troubled than human beings. It
is only human arrogance that assumes that it is always preferable to be a human
rather than a member of another species.

[2] The classic work on this topic is Lovejoy (1936); see also Bowler (2003).

both existence and life, but they also have desires and appetites and the capacity for self-generated motion. The pole position within the natural world is occupied by – no prizes for guessing – human beings. Humans possess all the attributes of the lower forms (existence, life, appetites, self-generated motion), but we also possess *reason*. This makes us particularly special, for it makes us more than mere matter. We are dual creatures, possessing material bodies but also immaterial minds or souls. Angels, in contrast, have no material component; they are pure spirit. As Saint Augustine put it, man is 'an intermediate being ... between beasts and angels'.[3]

There were further subdivisions within the Great Chain of Being. Among humans, for example, kings were considered higher on the scale than aristocrats, and aristocrats higher than peasants. Inevitably, some also suggested that the different human races occupied different ranks. Ideas such as these were optional extras, but the basic assumptions of the Great Chain of Being were – and still are – deeply entrenched in the Western view of the world. In summing up its influence, Arthur Lovejoy wrote that the Great Chain of Being 'has been one of the half-dozen most potent and persistent presuppositions in Western thought'.[4]

None of what I've said is meant to suggest that we've only ever had good things to say about ourselves. In the Christian tradition, people are natural-born sinners, and although we're superior to the animals, angels are superior to us. Nonetheless, even when we haven't explicitly emphasized just how wonderful we are, we've typically accorded ourselves a vastly important place in the world. In the monotheistic religions of the West, for example, we are the central interest of God, the being responsible for creating and directing the entire universe (not a trivial matter, then). We are the centrepiece and purpose of creation. The universe exists so that we can exist; the sun, moon, and

---

[3] Augustine (2003), p. 359.
[4] Lovejoy (1936), p. vii.

stars exist to shine upon us. Thus, although we have viewed ourselves as flawed and sinful, we have also believed that we occupy a unique and crucial position in the cosmos. This pair of beliefs sits in an uneasy tension in traditional Christian thought.

Pushed from the pedestal

It seems almost inevitable that evolutionary theory will have implications for the question of our status in the natural world. Numerous thinkers have pointed out that, like the Copernican revolution before it, the Darwinian revolution deflated human beings' view of their own importance in the grand scheme of things. Some even argue that Darwin's theory was a greater blow to our esteem as a species than the Copernican revolution. 'It is not just that we are on a speck of dust whirling around in the void', wrote the philosopher Michael Ruse, 'but that we ourselves are no more than transformed apes.'[5]

There are various ways that evolutionary theory might take us down a peg. For one thing, it recasts us as one species among countless millions. Not only that, but on an evolutionary timescale, we have been here for barely a blink of an eye. Traditionally, human history was viewed as coextensive with the history of the world. But once evolutionary theory removes the scales from our eyes, the human species is revealed as a new entrant on the stage of life. There was a time – a vast, gaping, incomprehensibly immense period of time – when human beings did not exist. This makes it hard to believe that humans or humanlike creatures could be the very purpose of creation. It stretches credulity to imagine that God instituted a process that would take nearly fourteen billion years solely for the purpose of creating a species that has been around for about two hundred thousand years and is unlikely to last for anything like another two hundred thousand. Why would God take such a costly and inefficient

---

[5] Ruse (1986), p. 274.

route to create a somewhat nasty vertebrate on a tiny planet in some lonely corner of the universe?

In other ways as well, evolutionary theory challenges the idea that God intended for us to be here. The variants that natural selection utilizes are not made to order but come about purely by chance. The chance element in evolution appears to undermine the notion that we are part of God's plan or that God intended for us to be here.[6] After Darwin, the idea that we are the centre of the universe, and that the world was made expressly for our habitation, seems quaint and embarrassingly anthropocentric. It is the species-level equivalent of solipsism. Evolutionary theory reverses the traditional view: the environment was not created for humans; humans evolved to fit the environment.[7]

It seems like an open-and-shut case: Darwin's theory recasts us as one species among millions, without any convincing claim to superiority. On the other hand, even among those who accept unreservedly that we evolved, not all accept that we are merely one among the animals. After all, even if we don't stand outside nature, and even if we're not quite as special as we'd once thought, this doesn't rule out the more modest claim that we are superior *within* nature and superior among the animals. One way to frame this view would be to suggest that the Great Chain of Being remains intact, but that evolutionary theory provides an updated account of the origin of the life forms found in this hierarchy. Through the unfolding of evolutionary history, we get the rank ordering of the Great Chain, with primitive bacteria leading to simple animals leading to more complex and intelligent animals, and culminating in human beings, the apex of the organic world, the crown of creation (at least for now).

---

[6] Of course, one could argue (indeed, many have argued) that the capriciousness of evolutionary history is its unpredictability *to us*, but that God is omniscient and would have known how things would come out. I'll leave it to the reader to assess this claim.

[7] More precisely, organisms and their environments co-evolve (Sterelny, 2003).

An early exponent of this type of view was the eighteenth-century French naturalist Jean-Baptiste Lamarck, the man who put forward the first real theory of evolution (see Chapter 2). Lamarck saw evolution as a progressive march up the Great Chain of Being, with humans the purpose and pinnacle of the evolutionary epic. In his view, the origin of new species is the product of a native impulse in life to advance into higher degrees of perfection and complexity – in other words, to inch its way up the ladder of progress. Lamarck held that all species were making an inevitable pilgrimage from inanimate matter to complex and intelligent selfhood, but that they had started at different times and some had therefore made more progress than others. Human beings belonged to the most ancient lineage and consequently were the most evolved species. Dung beetles were part of a more recent lineage and thus were less evolved.

It is now known that this view of evolution is false. All life on this planet can be traced back to a common ancestor, and thus has been evolving for precisely the same length of time. Furthermore, evolution is not the product of a native impulse in life to advance towards perfection; it is a product of the mechanistic and unguided process of natural selection. Nonetheless, it may be possible to link the Great Chain of Being with modern evolutionary theory. Here's how it could work. Natural selection produces ongoing and inevitable progress, starting with the simplest organisms and creating better and more complex designs as time goes by. Evolutionary change occurs because poorer designs are replaced with better ones. Today's organisms are superior to and more evolved than organisms that lived in earlier periods of time. Human are superior to dinosaurs, for example, because the process of replacing poorer with better has been going on for so much longer. For the same reason, humans are superior to our recent ancestor, *Homo erectus*. *Homo erectus* and other extinct species are evolution's losers; those alive today are its winners and are therefore superior. Or so it might be argued.

It might also be argued that, among existing organisms, some are superior to others. This is not because the superior ones belong to lineages that have been evolving for longer, as Lamarck thought, but because animals that attained their current form long ago are less evolved than more recent additions to the stable of life. Human beings are more evolved than crocodiles because crocodiles have barely changed since the age of the dinosaurs. They've travelled less distance along the preordained path of evolutionary progress. Using the same logic, one might conclude that mammals are superior to reptiles because they evolved from reptiles; that reptiles are superior to amphibians because they evolved from amphibians; and that amphibians are superior to fish because they evolved from fish. (Note that this scheme would not justify our intuition that we are superior to birds, as birds and mammals both evolved from reptiles.)

A related view is that evolution involves an overarching trend towards greater intelligence and/or complexity, and that human beings, as a recent species, are the most intelligent and complex products of natural selection to date. We are more evolved than chimpanzees because our lineage has taken more steps along the path of evolutionary progress, and we are therefore smarter and more sophisticated than they are. For the same reason, chimpanzees are more evolved than monkeys, and monkeys more evolved than snakes.

With ideas such as these, many early evolutionists thought they could rebuild the Great Chain of Being on a strictly Darwinian foundation. Sadly, this even extended to the racist aspects of the Chain. Some early theorists argued that certain human races were more evolved than others. The biologist Ernst Haeckel, for example, suggested that, among human beings, Europeans were more evolved than other racial groups, and that, among Europeans, Germans were the most evolved. (Try to guess what his nationality was.) Meanwhile, other thinkers looked not to the past but to the future. With his usual flare, Nietzsche observed that:

Formerly one sought the feeling of grandeur of man by pointing to his divine *origin*: this has now become a forbidden way, for at its portal stands the ape, together with other gruesome beasts, grinning knowingly as if to say: no further in this direction![8]

But now, Nietzsche noted disapprovingly, instead of pointing to man's *origin* for the feeling of grandeur, Darwin hints instead that 'the way mankind is *going* shall serve as proof of his grandeur and kinship with God'.[9] Many thinkers suggested that humans would continue to progress and improve over time, ultimately evolving into a superior species which, in Alexander Pope's words, would 'show a Newton as we show an ape'. Nietzsche himself came to the view that human beings are no more than a bridge between our pre-human ancestors and a superior species. 'Man', he wrote 'is something that is to be surpassed.'[10] Such ideas, like all those we've considered in this section, rest on the assumption that evolution is synonymous with progress. It's time to evaluate this assumption.

Does evolution imply progress?

The idea that evolution involves ongoing progress is widespread, especially among the lay public.[11] The best-recognized emblem of evolution starts with an ape, its knuckles dusting on the ground, and then moves through a series of intermediate forms to a fully upright human (a male human, to be exact). This is not usually viewed as a mere depiction of the historical sequence of events leading to our species; it is an emblem of progress. And from a certain perspective, it certainly does look as though there's been progress in evolution. In the beginning, this planet was host to nothing more sophisticated

[8] Nietzsche (1982), p. 32. [9] Nietzsche (1982), p. 32.
[10] Nietzsche (1993), p. 44.
[11] For good discussions of the issue of evolutionary progress, see Ayala (1988); Dawkins (1998a); Gould (1996); Mayr (2001); Nitecki (1988); Ruse (1997); Williams (1966).

than primitive bacteria. Now there is a huge diversity of life forms so complex and sophisticated they boggle the mind – and, for that matter, there are minds, including human minds, which are capable of understanding the fundamental laws of the physical world and the processes that brought life and mind into existence in the first place. Surely, then, there has been progress in evolution.

The first point to make – and we'll come back to this point again and again – is that although natural selection produces change, to label this change *progress* is to give it a positive valuation that is often difficult to justify. Natural selection does not always favour things we think are good. It favours any trait that enhances the propagation of the genes giving rise to it, regardless of whether we think it desirable morally, aesthetically, intellectually, or in any relation to any other standard by which human beings measure progress. An illustration: when Jim Morrison of rock band the Doors died at age twenty-seven, he supposedly had twenty paternity suits filed against him. The life-style that put him in this situation was maladaptive in the everyday sense of the word (because it caused him legal headaches, harmed other people, etc.), but was adaptive in the evolutionary sense (because he left more offspring than a less reckless person would have). This is an example of how natural selection does not necessarily favour traits that we consider good or healthy or desirable. I'll give you another: the course of evolution has furnished animals with an increasingly great capacity to suffer. Monkeys suffer more than trilobites, and humans presumably suffer more than monkeys. Evolution has also furnished animals with an increasingly great capacity to *inflict* suffering. As evolutionary history has unfolded, then, the universe has come to contain more and more suffering. Is that progress?

Here's the bottom line: evolution is defined as a change in the frequency of genes in a population. Nothing in this definition implies progress or betterment. If gene frequencies are changing, then regardless of the direction of change, evolution is taking place. It's as simple as that. But this conclusion has some surprising consequences. Some scientists suggest that, in modern civilizations, the unintelligent, lazy, and reckless are having more offspring than the intelligent, industrious, and conscientious, and that as a result, any genes predisposing people to the former traits are becoming relatively more common.[12] That is, on average, we as a species are becoming genetically less intelligent, less industrious, and less conscientious (see Chapter 12). This may or may not be true; my purpose in mentioning it is simply to raise the question: does this sound to you like evolution? If it doesn't, or if it sounds like the opposite of evolution, then you haven't fully grasped what evolution is. Evolution is change. It is not necessarily change we consider good.

On the other hand, we could always put our traditional standards of goodness aside and adopt the standards utilized by natural selection instead. We could say that whatever increases evolutionary success is good, and therefore that natural selection necessarily brings about progress. It's not clear to me why we'd want to do this, but even if we did, we still wouldn't get the outcome we're after. First of all, the criterion of evolutionary success wouldn't necessarily put us at the top of the hierarchy of life. In terms of sheer numbers, beetles are vastly more successful than we are. (As the biologist J. B. S. Haldane said, if there is a God, he has 'an inordinate fondness for beetles'.) We're not even the most successful of mammals; rats can claim that title. But rats and beetles are chickenfeed when you start looking at life at the microscopic level. The most common organisms are, and always have been, bacteria. In popular accounts of evolutionary history, the present age (the Cenozoic era) is known as the Age of Mammals; the time when dinosaurs walked the earth (the Mesozoic era) is known as the Age of Reptiles; and the preceding age (the Palaeozoic era) is known as the Age of Fish. But this is not an accurate way of construing the history of life on earth. As Stephen Jay Gould

---

[12] Lynn (1996).

noted, this is actually the Age of Bacteria, and it always has been.[13] Bacteria are the dominant and most numerous life forms on earth. Macroscopic organisms such as ourselves are a strange and rare aberration – a deviation away from the main trend in evolution, which is an increase in the number of bacteria.

There is a second and more fundamental problem with defining progress in terms of evolutionary success. Adaptations are only better or worse in the context of the environments in which they are found. As the environment changes, the criteria for goodness of design change with it. There is therefore no reason to think that evolution will produce ongoing progress, or that later designs will be superior to earlier ones. The fact that we evolved later than the dinosaurs, for instance, does not mean that we are better adapted than they were. Modern animals are better adapted to the modern environment than ancient animals would have been, but ancient animals were better adapted to *their* environment than modern animals would be. If dinosaurs were reanimated and thrown together with humans in a gladiatorial contest of survival, there is no guarantee that humans would come out on top. In fact, if the humans were put there in a technology-naked state, I'd put my money on the dinosaurs. Changes in the inhabitants of the earth do not reflect a constant process of improvement in the design of organisms, any more than changes in fashion over the years reflect a constant process of improvement in the quality of clothing.

This has a number of repercussions. One is to challenge the idea that extinction implies evolutionary failure. The dinosaurs are often tarred with this brush, but they were a hugely successful group of animals: they survived for 160 million years. Sure, they eventually died out. But this is the ultimate fate of all life, here or anywhere else in the universe. The fact that the dinosaurs are gone and that we're still around doesn't mean we're superior. It means that their time on the stage of life just happened to come sooner than ours did. It's easy

[13] Gould (1996).

to imagine a parallel universe in which humans came first and dinosaurs evolved after we went extinct, rather than the other way around. It's also easy to imagine a parallel universe in which humans never evolved at all. As Gould noted, if it weren't for the asteroid that smashed into the planet and wiped out the dinosaurs sixty-five million years ago, dinosaurs would still be the dominant form of life on earth today (ignoring microscopic life).[14] It wasn't until the dinosaurs were out of the picture that mammals were able to stage their takeover bid. If the dinosaurs *hadn't* been wiped out, most mammals would still be mouse-sized, nocturnal insect-eaters, and we would not be here. The reason mammals took over from the dinosaurs was not that they were superior; the dinosaurs just happened to be in the wrong ecological niche at the wrong time, and mammals were able to fill the gap left in their wake.

A similar point can be made with respect to extinct human species such as *Homo erectus*. We often assume that we're superior to this hominin species. But *Homo erectus* occupied the earth for around 1.6 million years. This is longer than we've been around, and longer than we're likely to be. Thus, in terms of the longevity of our species, we will probably have to be counted as evolutionary failures – except that there'll be no one around to do the counting. This raises an important point: intelligent beings can only contemplate these kinds of questions before they go extinct, and for that reason they may be tempted to view themselves as the endpoint of evolution. It is difficult to view our species in the context of the entire history of life on earth, for the simple reason that the entire history of life on earth hasn't happened yet. We mustn't forget, though, that for all our assets and achievements, we are just as vulnerable as the dinosaurs were. We could be wiped out at any time by a stray asteroid or an especially virulent virus (a David to our Goliath).

We must next challenge the whole notion of a prescribed path of evolutionary progress. People often talk about the course of evolution

[14] Gould (1996).

'from monad to man'. But this is no more sensible than talking about the course of evolution from monad to, say, anteater. Today's anteaters are no less and no more evolved than today's human beings, at least in the sense that humans and anteaters – and tigers and turtles – have been evolving for precisely the same length of time. It is not the case that some species are more evolved than others; instead, we've all just evolved in different directions. The fact that we possess traits such as reason, language, and morality, whereas other animals don't, is *not* because we are more evolved than they are, any more than the fact that some animals possess a superior sense of smell or a greater propensity towards violence means that *they* are more evolved than *us*. We are all equally evolved. If this idea doesn't sound right to you, then you're still mixing the concept of evolution with the pre-Darwinian notion that all life can be ranked from lowest to highest. Stop it!

As soon as we abandon the idea of a preordained path of evolutionary progress, we also have to abandon the idea that animals that attained their current form in the distant past (e.g., crocodiles and tuataras) are less evolved than those that appeared on the earth more recently (e.g., us). Indeed, one could just as well argue that the older species are *more* evolved, on the grounds that they hit upon a lifestyle that was more successful in the long term than those of any of the various species that led to *Homo sapiens*. I'm not actually advancing this view; my point is just that there is no obvious or self-evident way to rank species, and the standards we use when we try to do so are often transparently anthropocentric.

No evolutionary biologist today would accept that evolution involves an inexorable march up the Great Chain of Being. However, some still think it involves progress in a more limited sense. Specifically, they argue that there are certain large-scale trends in evolution.[15] The two most common claims of this nature are: (1) that there is an overall increase in the complexity of life; and (2) that there is an overall

[15] See, e.g., Wilson (1999).

increase in intelligence (from the amoeba to Einstein, as Karl Popper put it).

This might seem like a promising approach, but let's not get too excited just yet. To begin with, a strong case can be made that there are no such trends in evolution. Take complexity. Certainly, selection will favour an increase in complexity if this enhances fitness. However, it will also favour a *decrease* in complexity if *this* enhances fitness. The modern tape worm is much less complex than its ancestors. The ancestors had nervous systems and digestive systems, but somewhere along the line tape worms shed both of these things. Natural selection goes with whatever works best at the time. If a simpler option does a better job – as it often does – then so be it. As for the claim that there is a large-scale trend towards intelligence, well we know that humanlike intelligence did not evolve for nearly four billion years of life on earth, and there is no obvious reason why it could not have appeared sooner. This hardly suggests an inevitable progression towards intelligence. People sometimes say: 'Dinosaurs were around for millions of years. Why did they not evolve humanlike intelligence?' But for us to say this may be the equivalent of a bird saying: 'Humans have been around for hundreds of thousands of years. Why have they not evolved the power of flight?' If birds could think as well as humans (but no better), they might assume that the most evolved mammals are the bats, as bats are the only mammals with the capacity for flight. It is no more plausible to think that there is a large-scale trend towards intelligence than it is to think that there is a large-scale trend towards flight – or sense of smell or propensity for violence. Intelligence may appear periodically and then disappear again, replaced by something new in the ongoing process of evolutionary experimentation found on this planet.

Second, even if there *are* large-scale trends in evolution, we still have to ask whether it would make sense to label this 'progress'. This question is particularly pertinent when it comes to complexity, and to answer it we need to return to a point we've touched on already. The word *progress* has positive connotations, but how can we justify the idea that complexity is better than simplicity? Why not claim that life started in a superior state of *simplicity*, but then degenerated into complexity in some lines of descent? Evolutionary theory itself provides no reason to think that complexity is superior. It all depends on what we choose to value. Sometimes we value complexity above simplicity; often, though, we do the reverse. In science, for example, if two theories are equally good at explaining the same set of facts, scientists opt for the simpler one. The moral of the story is this: even if there *is* a large-scale evolutionary trend towards complexity (or anything else), there are no objective grounds to say that this is a good thing. If you like it, it's a good thing; if you don't, it's not. There is nothing else to say about it.

A third point is that even if there really are large-scale trends in evolution, it is not necessarily the case that humans will be at the forefront of these trends. Clearly, if there's a large-scale trend towards intelligence, we're at the forefront of that. But with respect to complexity, things are less clear. Are we really the most complex organism on the planet? We have the most complex brain and culture, but that doesn't mean we're the most complex in every significant sense.[16] Consider genetic complexity. People tend to assume that we are genetically more complex than any other animal. But most mammals have a genome similar in size to our own. Not only that, but the red viscacha rat (*Tympanoctomys barrerae*) has a genome more than twice as large as ours, and the marbled lungfish (*Protopterus aethiopicus*) has a genome around *forty times* as large.[17] When confronted with this information, people never think 'Oh, OK then, I guess we're less evolved than red viscacha rats and marbled lungfish.' Instead, they start scouting around for another standard against which to judge evolutionary progress, a standard that puts us at the top of the

[16] Williams (1966).
[17] Makalowski (2007).

hierarchy of life. People start with the conclusion that we're superior and then look for a way to justify it.

There is one final sense in which it might be said that evolution is progressive. Richard Dawkins defined progress as 'an increase, not in complexity, intelligence or some other anthropocentric value, but in the accumulating number of features contributing towards whatever adaptation the lineage in question exemplifies'.[18] Thus lions become progressively better hunters, whereas gazelles become progressively better escape artists. If, after an evolutionary arms race between lions and gazelles, you took a modern lion and sent it back in time to an earlier point in the arms race, the modern lion would outperform other lions and decimate the gazelles. Similarly, if you took a modern gazelle and sent it back in time, it would outperform other gazelles and escape the clutches of the fittest lions. In that sense, and within a narrowly defined context, later animals are better adapted than earlier ones. Dawkins noted that, 'By this definition, adaptive evolution is not just incidentally progressive, it is deeply, dyed-in-the wool, indispensably progressive.'

There is no denying that there are local trends in evolution. In fact, without local trends, there could *be* no evolution. However, there are two points to make. The first is that, even if we decide to call these local trends 'progress', progress in this sense is a far cry from the type of ongoing and absolute progress that might support claims of human superiority. There is no overall trend towards improvement. Instead, members of each species get better at doing whatever it is they do: lions get better at doing lion things; people get better at doing people things. If this is progress, then it occurs only within species, and cannot be used to rank species against one another. It also occurs only in relation to a specified set of selection pressures. As soon as the selection pressures change, the criteria for judging progress change as well. Thus, life progresses in certain directions for a while, then the

[18] Dawkins (1998a), p. 1017.

environment changes or a mass extinction wipes out much of the existing progress, and it all starts again, progressing in multiple directions for no particular reason.

That's the first point. The second point is that we must ask again whether 'progress' is the appropriate label to use. And the problem we keep coming back to is that the word has positive connotations, but local trends in evolution are not things we necessarily think are positive. If humans are becoming less intelligent, for example, that's a local trend, but it's not one we'd want to call progressive. Why not just talk about local trends in evolution? If talking about local trends in evolution seems to say less than talking about progress, then talking about progress says more than can be justified. Our overall conclusion is that we must reject the view that evolution is progressive.

The spark of reason

The upshot of this rather long discussion is that if we wish to maintain that human beings are superior among the animals, this cannot be based on the idea that evolution is progressive, and that humans are the latest and greatest of its products. However, it's an easy enough matter to decouple claims of human superiority from the concept of evolutionary progress. The way this is most often done is by pointing to our intellectual abilities – our unique possession of reason, language, and the capacity for culture. Claims of superiority based on traits such as these need not rely on evolution being progressive. But can they survive critical inspection?

We'll start with the most extreme version of this general line of thought: the claim that we and we alone are rational, whereas all other animals operate purely on instinct. An obvious criticism of this claim is that it is the product of an anthropocentric bias. In his book, *The Descent of Man*, Darwin made a strong case for the continuity of the mind and intellectual abilities of humans and non-humans.

> The difference in mind between man and the higher animals, great as it is, certainly is one of degree and not of kind. We have seen that the senses and intuitions, the various emotions and faculties, such as love, memory, attention, curiosity, imitation, reason, &c., of which man boasts, may be found in an incipient, or even sometimes in a well-developed condition, in the lower animals.[19]

Darwin didn't merely assert this; he provided many examples. Indeed, in keeping with his characteristic method, he positively bombarded the reader with examples. Subsequent research has only confirmed Darwin's suspicions. Other animals are not nearly as far behind us intellectually as we like to think. Chimpanzees, for example, are remarkably clever animals, capable even of such sophisticated cognitive activities as wilfully deceiving one another.[20] Whenever scientists observe that we're not so very different from other animals, a certain segment of the population views this as belittling our species. If nothing else, this reveals these people's attitudes about animals! But it doesn't change the facts, and the fact is that our intellectual talents are not quite as unprecedented as we've been led to believe.

Where did this notion that we alone are rational come from? The psychologists Michael Tomasello and Josep Call make an interesting suggestion:

> The Western intellectual tradition was created by people living on a continent with no other indigenous primates. It is therefore not surprising that for more than 2,000 years Western philosophers characterized human beings as utterly different from all other animals, especially with regard to mental capacities.[21]

Not only that, but the fact that there are no longer any Neanderthals or other human species on the planet must have encouraged people in all parts of the world to characterize *Homo sapiens* as very different

---

[19] Darwin (1871), p. 151.
[20] Whiten (1997).
[21] Tomasello and Call (1997), p. 3.

from other animals and as uniquely rational. We're not, though; other animals are smarter than we think.

How might a human supremacist respond to this? An immediate response would be that it doesn't rule out the possibility that we are superior; it implies only that we're not *as* superior as we once thought. Besides, there *are* some unique differences between humans and other animals. The most striking is the capacity for language. We know that some animals, such as chimpanzees and dolphins, can learn to use a handful of words. However, among extant species, the capacity for grammatical language is almost certainly unique to human beings. This sets us apart from other animals in an important way. Language allows knowledge to be transmitted from mind to mind, and to accumulate across generations. It therefore puts our species on a different plane of understanding from any other species. (It also puts us on a different plane of *mis*understanding from any other species – that is, we have a greater capacity than other animals to acquire sophisticated false beliefs as well as sophisticated true ones – but let's put that aside for now.) Language and culture have given us unprecedented new powers that have allowed us to assume the top dog position among the non-microscopic species on this planet. Should we conclude, then, that we are superior among the animals?

Let's first allow that we are indeed unique in many ways, language being the most important. This is all very well and good, but as the sociologist Pierre van den Berghe noted, 'we are not unique in being unique. Every species is unique and *evolved* its uniqueness in adaptation to its environment.'[22] Furthermore, we shouldn't be too over-awed by the differences between our intellectual powers and those of our animal cousins. Although we may be notably smarter than any other species alive today, it's not so long ago that we shared the planet with other human species, and we're not that much smarter than they were. There's a deeper problem as well. Think about this: humans and

[22] Van den Berghe (1990), p. 428.

other animals come in different shapes and sizes, but when compared to the size of the universe, we're all approximately the same size. On a scale spanning from the smallest particles to the entire visible universe, none of us is significantly different in size from an ant or a blue whale. The differences are important to us, but on a broader scale they fade into insignificance. In exactly the same way, we are vastly more intelligent than any other species on this planet, but this only seems important when we view things on a scale with the simplest organisms at one end and us at the other. If our scale extended instead to a hypothetical mind capable of comprehending the full complexity of the universe, our intellectual abilities would be far from impressive, and would not be significantly different from those of an ant or a blue whale. A mind that could accurately represent the universe in all its intricate detail would be unimaginably greater than our own. Far from having such powers ourselves, however, we are barely capable of understanding that tiny fragment of the universe which is the human brain (that is, our brains are barely capable of understanding themselves).

There's another point to make, and it's even more decisive. At the start of this section, we examined the claim that humans and humans alone possess reason, and that this makes us superior among the animals. It was then suggested that this claim was anthropocentric, on the grounds that other animals actually share some of the greatness bestowed by the possession of reason. But this suggestion is itself anthropocentric, for it is based on a tacit assumption that human reason can be taken as the universal standard against which to judge the ultimate worth of any animal. Neither the superiority nor the uniqueness of our intellectual abilities is relevant to the issue of our status among the animals. We've been sidetracked. Even if we outclassed other animals in all our intellectual abilities – indeed, even if all our intellectual abilities were utterly unique – this would not warrant the conclusion that this puts us above the other animals. The Darwinian perspective recasts reason as an adaptation, and in the process demystifies it. Thus, even if it were true that humans differ from all other animals in their intellectual abilities, this would have only the same significance as the fact that elephants differ from all other animals in possessing trunks. Certainly, humans are better at reasoning and long-range planning than any other animal. However, this is no more significant than the fact that lions are better at running than woodpeckers, or that woodpeckers are better at pecking than lions. We can legitimately say that we are superior to other animals *in certain traits*, our intellectual faculties being the most obvious. However, other animals are superior in other traits, and our intellectual superiority does not make us superior in a global sense.

Are humans inferior to other animals?

I'm not going to deny that it's possible to select criteria that make us superior among the animals. If we choose to judge animals in terms of their linguistic prowess, for instance, we'll obviously secure the number 1 ranking. The problem is, though, that it's *just as easy* to choose criteria that make us inferior. To underline this point, let's flip things on their head and ask a new question: are humans *inferior* to other animals? To get the ball rolling, we'll continue our discussion of intelligence or reason. I argued above that reason is merely one adaptation among many, and thus that it affords us no special claim to superiority. It might be objected that reason is superior among the adaptations. However, although our capacity for reason has been highly adaptive thus far (consider the size of the human population), in the longer term, it is quite plausible that our intelligence – and the technologies it makes available to us – could lead us to drive ourselves to extinction. Thus, if judged in terms of adaptiveness, or in terms of the long-term survival of the species, intelligence may turn out to be *inferior* to the adaptations found in other animals. If intelligent organisms have a habit of rapidly exterminating themselves, as many commentators suggest they might, then natural selection provides as much of a stamp of approval for the rapid extinction of intelligence as it does for the emergence of intelligence in the first place.

There are also various *moral* arguments for concluding that we are an inferior form of life. Evolutionary theory shows us that God did not put us in our position of pre-eminence in the world. We fought and clawed and bullied our way into it. To some extent, our success as a species is a mark of our despicability. Let's get specific. Human beings have a long and horrific record of genocide, and this is something that marks us out from the other animals. (For some strange reason, though, we seem to overlook this when contemplating what makes us unique.) In judging human history, we condemn those individuals who engage in genocide. But if we use *the same standard* to judge our species' relative worth within the animal kingdom, we would have to conclude that, in this respect, we are inferior to all other animals. Adopting other standards leads to the same conclusion. As humans have come to occupy every nook and cranny and corner of the globe, our arrival has coincided with the extinction of much of the flora and fauna in those regions.[23] It is not only European colonialists or industrialized societies that have ravaged the environment. Human beings are destructive animals, and as a result of our intelligence we are capable of destruction on a much greater scale than any other species. We condemn individuals who play fast and loose with the environment. But if we use the same standard to judge our relative worth among the animals, we would have to conclude again that we are inferior. The conclusion is as inevitable as it is uncomfortable: not only is it possible to choose criteria that make us inferior to other animals, when judged by criteria that *we ourselves routinely use*, we are inferior.

Not surprisingly, the idea that humans occupy the bottom rung on the scale of life is not a popular one. Yet various thinkers and misanthropes have toyed with it. According to Mark Twain, for example:

---

[23] Diamond (1992, 1997a).

Of all the creatures that were made, [man] is the most detestable. Of the entire brood he is the only one – the solitary one – that possesses malice ... He is the only creature that inflicts pain for sport, knowing it to *be* pain.[24]

More recently, the philosopher John Gray wrote that '*Homo sapiens* is only one of very many species, and not obviously worth preserving. Later or sooner, it will become extinct. When it is gone the Earth will recover.'[25] In Gray's view, the earth is presently suffering from a 'plague of humans', but this cannot be sustained for long. And maybe that's not a bad thing. Perhaps the world would be a better place without us in it. Perhaps we should conclude, along with the philosopher Bertrand Russell, that human beings are a mistake.

The universe would be sweeter and fresher without them. When the morning dew sparkles like diamonds in the rising sun of a September morning, there is beauty and exquisite purity in each blade of grass, and it is dreadful to think of this beauty being beheld by sinful eyes, which smirch its loveliness with their sordid and cruel ambitions.[26]

Of course, I don't seriously want to argue that we are inferior among the animals. My serious argument is this: in order to assess our status in nature, we need to establish some standard against which to judge different species. In one sense, we can choose any standard we like, and some choices will put us on top. However, if we wish to argue that our choice is based on more than just an anthropocentric bias, we must show that it has some objective justification. The problem is that, in a Darwinian universe, this is not possible even in principle. One of the central elements in Darwin's theory is that there is no intention or intelligence behind the design found in the biosphere (excepting those things designed by evolved intelligent organisms). This implies that there is no ultimate justification for choosing any particular

[24] Twain (1924), p. 7.
[25] Gray (2002), p. 151.
[26] Russell (1953), pp. 48–9.

standard for judging life forms. Whereas, on a pre-Darwinian view, it was believed that our standards could be grounded in the intentions of God (or the will of nature or the intelligence of the universe), in a Darwinian world, our standards have no anchor. Any standard we choose will ultimately be arbitrary, and will have no more justification than any other – including standards that put us at the bottom of the heap.

Conclusion

Many Victorians were shocked by the supposed implication of evolutionary theory that human beings are descended from lower animals. We have now seen that the implications of Darwin's theory are far deeper and more shocking than this. It is not correct to say that humans are descended from lower animals. From a Darwinian perspective, the designation *lower* is without foundation. This is not to argue that all animals are equal, for that would be just another way of ranking species (i.e., giving them all the same rank). The argument is that any such ranking is always and necessarily misguided. Evolutionary theory does not provide an alternative explanation for the rank ordering found in the Great Chain of Being. Evolutionary theory undermines the possibility that there could even *be* a Great Chain of Being.

Thus we may conclude after all that we are one species among many, with no claim to special status. When viewed in this light, our previous convictions about the world and our place in it begin to seem untenable, to put it politely. The idea that the universe was created by a humanlike being (God), and that human beings are the purpose and centrepiece of creation, should be taken precisely as seriously as the view that the universe was created by, say, a wombat-like being, and that wombats are the purpose and centre of creation. Similarly, for humans to believe that intelligence is the crowning achievement of evolution should be given the same respect we would give a herd of elephants that reached the same conclusion in regard to their trunks. After Darwin, the idea that we are superior among the animals comes to resemble the overconfident bluster of a brash adolescent. If our species is ever to grow into mature adulthood, we must adopt a more realistic and sober appraisal of our true standing in the natural world.

TEN

Meaning of life, RIP?

Man is the result of a purposeless and natural process that did not have
him in mind.

George Gaylord Simpson (1967), pp. 344–5

You came from nothing and you return to nothing, so what have you
lost? Nothing.

'Always Look on the Bright Side of Life',
from the Monty Python film, The Meaning of Life

LISA SIMPSON: Maybe there is no moral to this story, Mom.
HOMER SIMPSON: Exactly! It's just a bunch of stuff that happened.

The Simpsons, 'Blood Feud'

What's the point?

This chapter is about the meaning of life and the purpose of the
universe.[1] What is the point of it all? *Is* there any point? Does it really
matter if there's not? Even the most successful among us may be
plagued by nagging doubts about the meaning and purpose of

---

[1] For some interesting works on this topic, see, e.g., Baggini (2004); Cottingham
(2002); Klemke (2000); Maisel (2009).

existence. Even for those who achieve wealth, fame, and acclaim beyond their wildest expectations – perhaps especially for these people – the question creeps back like an unwelcome visitor: what does it all mean? Is life no more than a downhill slide into the grave, in which we strive merely to postpone our deaths and prolong a pointless and uncomfortable existence?

These questions are probably more pressing for people living today than they were before the rise of civilization and science. When people's conception of the universe was limited to the small corner of the world they inhabited, it was easy to believe that they were a significant part of that universe. But as societies have grown and expanded, and as the vast size and age of the universe have become apparent, it has become harder and harder to believe that we have any real importance in the grand scheme of things. Science tells us that we are inconsequential specks of dust scrabbling around blindly on a pale blue dot orbiting a tiny star in an inconceivably large universe. Some say that size doesn't matter, but we all know that really it does.[2] How can anything we do be important?

It's not only cosmology that alters people's perceptions of their significance in the universe and the meaningfulness of life. Evolutionary theory threatens to do the same thing. When people contemplate the meaning of life in the light of evolution, they typically take one of two paths. They assume either that: (1) evolutionary theory tells us that the purpose of life is to transmit as many of

---

[2] To be fair, some interesting things have been written about why size *doesn't* matter. Archbishop William Temple, for example, noted that: 'I am greater than the stars for I know that they are up there and they do not know that I am down here.' And the mathematician and philosopher Frank Ramsey wrote:

> I don't feel the least humble before the vastness of the heavens. The stars may be large, but they cannot think or love; and these are qualities which impress me far more than size does … My picture of the world is drawn in perspective, and not like a model to scale. The foreground is occupied by human beings and the stars are all as small as threepenny bits.
>
> (Ramsey, 2001, p. 291)

our genes as possible to the next generation, or that (2) evolutionary theory implies that life is ultimately meaningless. I don't want to give away the punch line of the chapter too soon, but let me just say that my conclusion will be that one of these two assumptions is correct and the other false. Let's get started!

Wisdom of the ages

The meaning and purpose of life is a topic of perennial interest to human beings, and there have been many attempts to solve this ancient riddle. The main source of solutions has been religion. Among the views associated with the Western religions are these: that the purpose of life is to serve or submit to God; to know and love God; to love one's neighbour; to gain entrance to heaven; to overcome evil; to convert other people to one's religion; or to look after the planet.[3] Eastern views include the idea that the purpose of life is to break free of the cycle of reincarnation and karma, or to achieve enlightenment and be extinguished as an individual conscious entity. There are also various secular or religion-neutral answers to the question of life's meaning. The following quotations provide a representative sample of the sort of ideas that are out there (I particularly like the fourth one):

1. 'We are here to help each other get through this thing, whatever it is' – Kurt Vonnegut's son.
2. 'The purpose of our lives is to be happy' – the Dalai Lama.
3. 'The purpose of life is not to be happy. It is to be useful, to be honorable, to be compassionate, to have it make some difference that you have lived and lived well' – Ralph Waldo Emerson.
4. 'The true meaning of life is to plant trees, under whose shade you do not expect to sit' – Nelson Henderson.
5. 'You come into the world with nothing, and the purpose of your life is to make something out of nothing' – H. L. Mencken.

[3] See, e.g., Warren (2002).

6 'It's nothing very special. Try to be nice to people, avoid eating fat, read a good book every now and then, get some walking in, and try to live together in peace and harmony with people of all creeds and nations' – Monty Python.

Even before we start looking at evolutionary theory, it's easy to find holes and flaws in these suggestions. Starting with the religious suggestions, how can we be sure that a universe created by God has any ultimate purpose? Surely it's possible that an omnipotent God created a purposeless universe. Second, even if there *is* a purpose, and even if we're capable of discovering what it is, how do we know we'll find it satisfactory? After all, we can always ask: what is the purpose of this purpose? If the goal of life is to care for the planet, for example, we can still ask: what is the purpose of caring for the planet? Indeed, given that God supposedly created the planet in the first place, isn't this a little like creating busywork to keep children occupied? When we look at it this way, caring for the planet would seem to have less point if there *is* a God than if there isn't! There are other problems as well. If the purpose of life is to help people, what happens if we achieve a Utopia in which no one needs any significant help? Would life no longer have any meaning? Similarly, if the purpose of life is to get to heaven, or to convert others to the cause, what happens when these goals are obtained? Presumably life would cease to have any purpose or point. Perhaps we'll all end up sitting around in heaven agonizing over the question 'What is the point of this everlasting life?' And regardless of what the purpose of our existence might be, there's always another question we might ponder: what is the purpose of *God's* existence?

So that's the situation before we bring evolutionary theory into the picture. Evolution just compounds the problem. First, to the extent that the theory challenges the existence of God, it undermines every God-based answer to the question of the meaning of life. Furthermore, as we've seen, evolutionary psychology poses a challenge to the belief that life continues after death. Thus, an evolutionary perspective undermines any purpose that involves the survival of the mind outside the body. This includes such suggestions as that the purpose of life is to get to heaven or to escape the cycle of karma and reincarnation.

In a sense, though, all these criticisms are moot, because the implications of Darwin's theory cut much deeper than this. The theory forces us to rethink the very question of life's meaning and purpose. We come now to the main part of the chapter.

Evolution and the meaning of life

I opened this book with the question 'Why are we here?' The answer I gave is that we are here because we evolved. I'm fully aware that a lot of people – and I'm thinking not only of Creationists but of those who fully accept evolutionary theory – are going to feel that this is an unsatisfactory answer to the question. Certainly, it provides a good answer to *one interpretation* of the question. But this is not the interpretation that most people have in mind when they reflect on the issue of why we are here. The answer seems to sidestep the question as it is usually intended. The real intention of the question is probably better captured with an alternative question: 'For what *purpose* are we here?' What I'd like to argue now is that evolutionary theory does also provide an answer to this sense of the question.

I should start by making clear what I'm *not* going to argue. I'm not going to argue that evolutionary theory implies that the meaning of life is to survive and reproduce, or to enhance our inclusive fitness, or to put forward our genes, or anything like that. I mention this because a lot of people seem to think that it implies exactly that. They seem to think that, if we're evolved machines 'designed' to replicate ourselves or propagate our genes, then that's what we should do and that's how we should measure our success in life. However, the fact that we evolved through natural selection does not imply that our purpose must be to propagate our DNA. Evolutionary theory tells us where we came from, not what we should do now that we're here. (If you're not persuaded by this, skip ahead to Chapter 12 before reading any further.)

Nonetheless, evolutionary theory does have implications for the question of the meaning of life. To see how, we first need to cover some background ideas. Traditional explanations for the design found in organisms (e.g., the design found in the human eye) involve a style of explanation known as *teleological explanation*. Teleological explanations are framed in terms of purposes and future consequences. For example, we might say that the giraffe has a long neck *for the purpose* of feeding on leaves high in the trees. However, from a Darwinian perspective, this is actually the wrong answer. In fact, it's not just the wrong answer; it's the wrong *kind* of answer to questions in biology. The giraffe does not have a long neck to achieve this or any other purpose. It has a long neck because long-necked giraffes in the past were more likely to survive and reproduce than their shorter-necked counterparts, and thus long-necked giraffes were more likely to pass on the genetic coding for their long necks. This point is crucial to a proper understanding of evolutionary theory: the theory replaces teleological explanations in biology with historical explanations. Thus there is no teleological explanation for the giraffe's long neck, or for any other adaptation. There is only a historical explanation. To pre-empt an inevitable criticism, I'll concede right away that biologists *do* often talk about adaptations in terms of their purposes. But in such cases the teleological explanation is always shorthand for what, at bottom, is a historical explanation.

Let's return, then, to the question of why we are here. The point I'm going to make is probably already clear. We have considered various suggestions about why we are here – to gain entrance to heaven, to achieve enlightenment, to be happy, to make other people happy, to propagate our genes. These are all teleological answers. From an evolutionary perspective, they are not simply wrong answers to the question of why we are here; they are the wrong *kind* of answer. Darwin showed us that there is no reason to assume that any aspect of life was designed with a future purpose in mind. As such, there is no reason to think that there is a teleological answer to the question of why we are here; there is only a historical one. Thus, evolution provides answers to both senses of the question of why we are here, the historical and the teleological: *we are here because we evolved, but we are not here for any purpose.*

Naturally enough, not everyone agrees with this conclusion. The philosopher of science, Karl Popper, wrote:

> The success of Darwinian natural selection in showing that the *purpose or end* which an organ like the eye seems to serve may be only apparent has been misinterpreted as the nihilistic doctrine that all purpose is only apparent purpose, and that there cannot be any end or purpose or meaning or task in our life.[4]

Popper was certainly right about one thing: we can have ends and purposes and tasks in our lives, and evolutionary theory could never tell us otherwise. But that doesn't challenge the point I'm making. I'm not denying that we all choose little goals for ourselves, and that this can make our lives meaningful in the emotional sense of the term. However, if we're interested in the question of whether life is *ultimately* meaningful, as opposed to whether it's potentially emotionally meaningful, well, after Darwin, there is no reason to suppose that it is. There is no reason to suppose that life has any ultimate meaning or purpose.

Living without meaning

This might sound like a gloomy conclusion, especially for those who were brought up believing that the universe does have some overarching purpose or our lives some ultimate meaning. An initial point to make about this is that, even if it *is* a gloomy conclusion, this says absolutely nothing about whether it is a true or an accurate conclusion. The second point is that, as it happens, it's not necessarily a

---

[4] Popper (1987), p. 141.

gloomy conclusion at all. People often muddle up the question of ultimate purpose with the question of whether life is worth living. These are very different things. There's an important distinction between the *idea* that life is ultimately meaningless (which is an abstract, philosophical conclusion), and the *feeling* that one's own life is meaningless (which is a symptom of depression). Most people can live perfectly good and happy lives even while accepting at an intellectual level that life has no ultimate meaning, at least once they get used to the idea. Some even cheerfully accept the fact that life is meaningless and view it as amusing in a strange kind of way – a cosmic joke but without a joke teller. Perhaps some people recognize that it was simply a childish mistake to assume that the universe *does* have an ultimate purpose, and that we can get on with our lives without having to continue making this mistake. There is no logical contradiction in the idea that life is good but meaningless, and the universe awesome but ultimately pointless.

This is an issue that the existentialist philosophers of the twentieth century grappled with and agonized over, and many of them came to the same conclusion that I have: that in the final analysis, life is meaningless. But many found a silver lining in this cloud. A recurring theme in existentialist philosophy is that, if there is no meaning or purpose imposed on us from outside, then we're in the position where we're free to choose the purposes we have in our lives, both as individuals and as a species. For a lot of people, this is a deeply liberating idea. As E. D. Klemke wrote:

> An objective meaning – that is, one which is inherent within the universe or dependent upon external agencies – would, frankly, leave me cold. It would not be *mine* ... I, for one, am *glad* that the universe has no meaning, for thereby is *man all the more glorious*. I willingly accept the fact that external meaning is non-existent ... for this leaves me free to *forge my own meaning*.[5]

5 Klemke (2000), p. 195.

Will this response to the problem of meaninglessness be satisfactory for everyone? Almost certainly not. The idea that the meaning of life is just whatever we want it to be may seem glib and altogether too easy. Purposes chosen by God, or otherwise imprinted into the fabric of existence, seem so much more important than those that we choose for ourselves. And although in a meaningless universe we are free to choose our own purposes, we shouldn't duck the uncomfortable conclusion that these purposes are themselves ultimately purposeless, and that life is ultimately meaningless. Nonetheless, for those who can recalibrate their expectations a little, the notion that we can choose our own meanings and purposes may be enough. One could even argue that our creative endeavours and achievements and small acts of kindness are all the more impressive against the backdrop of a purposeless universe. Why can we not appreciate pointless kindness and beauty just as much (or even more) than we would appreciate kindness and beauty that somehow have an ultimate purpose?

Rejoining nature

One last point before we wrap up this chapter and this section of the book. It is often said that human beings have purposes but that the universe does not. This is not quite correct, however, for we must take into account our earlier conclusion (Chapter 8) that evolutionary theory undermines mind–matter dualism, and recasts the mind as a part of the physical universe. What this implies is that, if you have a purpose – a goal or orientation for your life – then a small part of the universe *has a purpose*. Our purposes are the purposes of the universe. The universe doesn't have *a* purpose, it has many – namely, those purposes possessed by the minuscule parts of the universe we call 'human beings' (and of course many other animals as well). This is not the grand purpose that we might have been hankering after, and we must remember, first, that there is no greater purpose behind these little purposes and goals (they are purposeless purposes), and, second, that natural selection itself has no purpose. Still, in a very real sense, it is false to say that the universe is purposeless. It *was* purposeless before the first life forms with purposes and drives evolved, and it will be devoid of purpose once more when the last life form takes its final gasp of breath. However, as long as we're here to contemplate such matters, to struggle and strive, the universe is not without purpose.

Conclusion

For better or for worse, human beings are unique among the animals in being able to comprehend the ultimate meaninglessness of existence. The philosopher Herman Tennesen provided a vivid, though perhaps somewhat overblown, description of our situation:

> Man's *in principle* unlimited capacity for insights and fore-sights – his ability to find out how things around him and in him operate, which is undoubtedly responsible for his present privileged place in the scheme of organic creation – this capacity, when extrapolated beyond its biological efficacy, may yield the most horrifying, vertiginously pernicious, unendurable insights into the fatuous futility and monstrous absurdity of … human existence.[6]

People want to think that there is some ultimate purpose or meaning behind their lives. Most probably there is not. Like the search for God, the search for the meaning of life is a wild goose chase. Darwin showed us that there is no reason to think that there is a teleological explanation for life. We are here because we evolved, and evolution occurred for no particular reason. Thus, on a Darwinian view, not only is our species not as special as we had once thought, but our lives are ultimately without purpose or meaning. Life just winds on aimlessly, a pointless, meandering sequence of events. Sometimes it's pleasant, sometimes not, but it lacks any overall purpose or goal or destination.

[6] Tennesen (1973), p. 408.

I don't deny for a minute that many of the standard suggestions about how to live a meaningful and purposeful life are beyond reproach. We need to remember, though, that we choose these meanings and purposes for our lives. They are not imposed on us from any external source. It's nice to make people happy and to help each other through this life. But these nice things are not the objective purpose of life; they're just nice things. If we decide, say, to plant trees under whose shade we do not expect to sit, that's our choice; it's not the meaning or purpose of life in any deeper sense than that. The only purposes the universe has are the little purposes found in those tiny parts of the universe that are complex, goal-driven animals. The universe has no overarching purpose, and most of its parts have no purpose at all. And in the end, it doesn't matter.

We've seen in the last few chapters that evolutionary theory represents a radical challenge to the pre-Darwinian self-portrait of our species, and to earlier views of our place in the world and the purpose of our lives. One possible criticism of the perspective I've presented here is that, in many ways, it is far from flattering. Leaving aside the obvious rejoinder – that this has absolutely no bearing on the issue of the accuracy of this perspective – I wish to wrap up this section by arguing that any such worries are misplaced. Indeed, the fact that this view is unflattering may be an *advantage* of an evolutionary perspective. Even a cursory glance at the track record of our species reveals that some of our most heinous acts have been committed under the intoxicating spell of our own sense of self-importance and righteousness. This suggests that a good dose of humility is probably in the best interests of us all – and when I say 'all' I refer not just to members of our own species but to sentient life in general. For a species inclined to see itself as the very purpose of the universe, some of the implications of evolutionary theory may be unpalatable. But we cannot infer from this that the ultimate effects of Darwin's intellectual revolution will be undesirable.

PART III

Morality stripped of superstition

ELEVEN

Evolving good

Scientists and humanists should consider together the possibility that the time has come for ethics to be removed temporarily from the hands of the philosophers and biologized.

E. O. Wilson (1975), p. 562

We believe that we should love our neighbour as ourselves, because it is in our biological interests to do so.

Michael Ruse (1988), p. 74

Morality is the device of an animal of exceptional cognitive complexity, pursuing its interests in an exceptionally complex universe.

Martin Daly and Margo Wilson (1988), p. 254

Evolutionary ethics

Evolutionary theory is unique in its ability to inflame passions and catalyse debate. One particularly heated area of debate concerns the impact of the theory for ethics. Some flatly deny that evolution has any relevance to ethics, a view they support by arguing that values cannot be derived from mere facts about the world. Others dispute this, although among this group there is little consensus about what exactly the moral implications of Darwin's theory might be.

Some argue that the theory supports laissez-faire social policies and the abolition of social welfare. Others argue that it forces us to rethink and recalibrate our moral priorities, and to reassess the value we place on the lives of humans v. other animals. Finally there are those who draw a darker conclusion, suggesting that the truth of evolutionary theory undermines morality altogether. In the remaining chapters, we'll consider each of these views in turn.[1]

Before we do that, though, we'll examine the evolutionary underpinnings of morality. The first thing I want to emphasize is that I'm *not* going to argue that morality is a direct product of biological evolution. I want to stress this at the outset because it's a point that has escaped some readers of my earlier writings on this topic. Several have complained that, although some of the *attitudes* embodied in our shared or formalized morality might have an evolutionary origin, formalized morality itself does not. This has been offered as an alternative to my position, but in fact it *is* my position. In this chapter, I argue that our moral codes are subject to influences other than those traceable directly to human nature or natural selection. This is not a mere qualification to my real thesis (supposedly that morality is a direct product of natural selection); it is a point that I consider central to the understanding of morality.

The moral animal

We all hold beliefs about which actions are right and which are wrong. Where do these come from? There are three main answers to this question. The first is that the mind directly perceives eternal and objective truths about rightness and wrongness. The second is that morality is a cultural invention and that we soak up our moral beliefs from the surrounding culture. A third possibility was opened up by Darwin's theory of evolution by natural selection. Starting with

[1] For an accessible and incisive discussion of this topic, see Kitcher (2006).

Darwin himself, various thinkers have argued that human morality has an evolutionary origin.[2] Like any other complex phenomenon in the biosphere, it is a product of evolutionary processes. When Darwin first put forward this view, it was, in many quarters, a profoundly unpopular suggestion. Few Victorians wanted to hear that their moral faculty, assumed at that time to be a gift from God, was instead an evolved adaptation. Darwin's wife Emma typified this reaction when she told their son Francis: 'Your father's opinion that all morality has grown up by evolution is painful to me.'[3] Since Darwin's day, the idea has continued to attract controversy. Nonetheless, there are good reasons to think that evolutionary theory can inform our understanding of the origins of the moral faculty. More than 100 years after Darwin put forward his views on the topic, the sociobiologist E. O. Wilson famously argued for the necessity of an evolutionary approach to morality:

> The emotional control centers in the hypothalamus and limbic system of the brain . . . flood our consciousness with all the emotions – hate, love, guilt, fear, and others – that are consulted by ethical philosophers who wish to intuit the standards of good and evil. What, we are then compelled to ask, made the hypothalamus and limbic system? They evolved by natural selection. That simple biological statement must be pursued to explain ethics and ethical philosophers, if not epistemology and epistemologists, at all depths.[4]

Admittedly, we could quibble with Wilson about various details: our emotional reactions are shaped in part by experience, and our moral values are not shaped only by our emotional reactions – we are taught values by our parents and other members of society, and sometimes, occasionally, we may even reason our way to an ethical conclusion. However, the core point stands: human moral systems are informed

---

[2] Darwin (1871). See also Alexander (1987); Axelrod (1984); de Waal (2006); Haidt (2001); Hauser (2006); Krebs (1998); Ridley (1996); Trivers (1971, 1985).
[3] Cited in Brooke (2003), p. 202.   [4] Wilson (1975), p. 3.

by emotions and preferences that have their origin in our evolutionary history. When philosophers consult their moral intuitions, they don't draw on some mysterious faculty connecting them with eternal moral truths. They draw on emotions crafted by natural selection.

People often struggle to come to grips with the full implications of this idea. For many, it is almost impossible to imagine that our attitudes on moral issues such as altruism, loyalty, infidelity, and the value of human life could be otherwise, and thus it is hard to believe that there is any need to invoke evolutionary theory or any other theory to explain these attitudes. Wilson has speculated that, if the capacity for morality had evolved among termites, these creatures would feel exactly the same way. However, their list of moral imperatives would differ somewhat from our own. Termite morality might emphasize:

> The centrality of colony life amidst a richness of war and trade among colonies; the sanctity of the physiological caste system; the evil of personal reproduction by worker castes; the mystery of deep love for reproductive siblings, which turns to hatred the instant they mate; rejection of the evil of personal rights; the infinite aesthetic pleasures of pheromonal song; the aesthetic pleasure of eating from nestmates' anuses after the shedding of the skin; the joy of cannibalism and surrender of the body for consumption when sick or injured (it is more blessed to be eaten than to eat); and much more.[5]

Viewed in this light, human moral beliefs begin to look less like reflections of timeless moral truths, and more like a set of principles fitting one particular species to its evolved lifestyle.

This is not to suggest that morality has no precedents in other animals; it clearly does. Non-human animals don't reason explicitly about right and wrong, but they do exhibit some aspects of human morality. Rather than being locked into an eternal war of all-against-all,

---

[5] Wilson (1996), pp. 97–9.

many animals display tendencies that we count among our most noble: they cooperate; they help one another; they share resources; they love their offspring. For those who doubt that human morality has evolutionary underpinnings, the existence of these 'noble' traits in other animals poses a serious challenge. When speaking of other species, we inevitably explain these traits in evolutionary terms. No one would want to explain the fact that female dogs love and care for their puppies as an arbitrary product of canine culture, for example. Given that we accept an evolutionary explanation for this behaviour in other species, it seems tenuous to argue that the same behaviour in human beings is entirely a product of a completely different cause: learning or culture. In principle, it is possible. However, we should have a strong reason to make this exception. Without such a reason, our default assumption should be that we are continuous with the rest of nature and thus that the behaviour has an evolutionary origin.

However, the fact that some aspects of moral behaviour have an evolutionary origin does not imply that morality as a whole is an evolutionary product. An immediate and very reasonable criticism of the idea that evolution accounts for all of morality is that human moral codes vary a great deal between different cultures and subcultures. Furthermore, moral beliefs sometimes change rapidly, both within individuals and within societies (take, for instance, the change in expressed attitudes about premarital sex in the Western world in the second half of the twentieth century). If humans shared a common moral psychology, there would presumably be little debate on ethical issues. Why, then, do we constantly wrangle over what's right and wrong? That said, we mustn't overestimate the variability in people's moral beliefs. There is a common core to all the moral systems of the world; there are, after all, no cultures in which dishonesty and cruelty are prized and honesty and kindness scorned, or in which courage is reviled and cowardice revered. Furthermore, professed moral beliefs vary more than our actual behaviour – in other words, although our moral beliefs sometimes stray from what evolution would prescribe if it had complete control, our behaviour doesn't stray quite as far. Nonetheless, there *is* considerable variation in people's moral beliefs, and plenty in their moral behaviour, and this makes it impossible to believe that morality is purely an evolutionary product, even if the imprint of our evolutionary history can be detected in our moral systems. To some degree, our codes of morality have a life of their own, free of evolved dispositions.

So there are legitimate criticisms of the idea that morality is wholly an evolutionary product. However, some commentators go too far in the opposite direction and end up making the opposite mistake, denying that evolution can shed any light at all on morality. Admittedly, at first glance, there appears to be a very good reason to do this: morality just doesn't seem to aid its bearers! Traditional moral systems tell us we should not act to satisfy our own selfish needs, but instead should act altruistically (that is, we should act to satisfy the selfish needs of other people). This hardly sounds like a recipe for evolutionary success. Our first task, then, is to consider how evolutionary principles might illuminate our understanding of this central component of morality: selflessness. We'll see that, although at first selflessness seems contrary to a Darwinian analysis, it is in fact an area of moral behaviour that evolutionary principles have shed a great deal of light on. It's not the only area, though; evolutionary theory has also shed light on sexual morality, and especially on our moral attitudes towards incest. In the remainder of this chapter, we'll consider each of these areas in turn, and then take a closer look at the role of cultural forces in the construction of our shared morality.

The problem of altruism

Darwin's theory of evolution solves one of the great mysteries facing the human mind: the origin and diversity of life on this planet. But, like any good theory, it also awakens us to new mysteries, taking aspects of life that seem so familiar as to demand no special explanation and transforming them into unsolved problems. Altruism is one example. In everyday life, the fact that people sometimes help one another hardly seems mysterious, especially when the recipients of that help are the helper's children or other loved ones. Indeed, it is often the *failure* to help that attracts people's attention.[6] Among evolutionary biologists, however, the important question is not why people sometimes fail to help, but rather why people (and other animals) *ever* help one another. E. O. Wilson described this as the central theoretical problem of sociobiology.[7]

The reason altruism is problematic is simple: if evolution is the product of a struggle between individuals to survive and reproduce, then presumably natural selection will help those that help themselves, rather than those that help others. The only organisms that would triumph in a competitive Darwinian world would be those that looked after number 1. Nonetheless, biologists have documented numerous examples of altruism. Typically, an altruistic act is defined as any act that involves a personal fitness cost to the actor but which enhances the fitness of the recipient. Acts meeting this description are found across the length and breadth of the animal kingdom, and even in plants. And that's a mystery. It's not what we'd expect.

The most extreme examples of biological altruism are found among a group of insects known as the *eusocial insects*. This includes ants, bees, and wasps. In each eusocial species, there is a caste of individuals that doesn't personally reproduce, but which instead helps the minority of reproductive members to do so. On the face of it, it is difficult to see how such behaviour could evolve. Given that sterile members of the group never reproduce, how is it that the genes contributing to their behaviour are maintained in the gene pool? And although

---

[6] Case in point: a famous book in social psychology (Latané and Darley, 1970) was entitled *The unresponsive bystander: why doesn't he help?*

[7] Wilson (1975).

eusociality represents an extreme of self-sacrificial altruism, lesser examples are ubiquitous. In various species, including Belding's ground squirrels, individuals risk their own lives by emitting a warning call when a predator is in the vicinity. The call alerts other individuals to the danger, and increases their likelihood of escaping unscathed. However, it also draws attention to the caller, increasing the caller's likelihood of falling prey to the predator. One might reasonably expect that callers would have fewer offspring on average than non-callers, and therefore that the tendency to issue alarm calls would be selected against. Yet it persists. How?

Various solutions have been proposed, and it is generally agreed that altruism is the product of multiple selection pressures. One of the most important pieces of the puzzle was the brainchild of a young graduate student, William D. Hamilton.[8] In 1964, Hamilton published the results of his doctoral thesis, a theoretical project on the origins of altruism. Hamilton had completed his thesis in virtual isolation, and his work had not met with the approval of his supervisors. There was even some doubt about whether he would get his thesis approved. This turned out to be highly ironic. Hamilton's article outlined a theory, soon dubbed *kin selection theory*,[9] which the biologist Robert Trivers later described as 'the most important advance in evolutionary theory since Darwin'.[10] E. O. Wilson went further, describing it as 'the most important idea of all'.[11] The theory helps to explain a particular type of altruism found throughout the animal kingdom, namely, altruism among genetic relatives. The basic idea is that, in certain circumstances, a gene 'for' kin-directed altruism can be selected (that is, increase in frequency in the population), for the simple reason that kin are more likely than chance to possess copies of the same gene. Relatives won't *necessarily* possess the gene,

---

[8] Hamilton (1963, 1964).
[9] Maynard-Smith (1964).
[10] Trivers (1985), p. 47.
[11] Wilson (1996), p. 315.

and non-relatives may, but a relative is more likely than a randomly chosen non-relative to possess it, and that's why kin-directed altruism can be selected.

Hamilton's theory provided an elegant solution to the problem of eusociality. It turns out that sterile eusocial insects, such as ants, are closely related to the reproductive members of their colony. As such, the genes underlying their altruism are passed on, not by the altruists themselves, but by their actively reproducing kin. As for Belding's ground squirrels, it turns out that they're most likely to issue warning calls when relatives are in the vicinity, and that this is especially true when there are *close* relatives in the vicinity.[12] Kin selection helps to explain much of the altruistic behaviour found throughout the natural world. It helps to explain why, in so many species, individuals help kin more than non-kin, and close kin more than more distantly related kin. And it helps to explain why human beings are no exception to this general rule. Like many other animals, our willingness to help tends to increase in step with our relatedness to the recipient of that help.[13]

This is a good start but there's an obvious limitation: not all altruism is directed towards kin. Altruism among non-relatives has been observed in various non-human species, from vervet monkeys to vampire bats.[14] And of course it is particularly common among humans. Thus, we're left with a new problem: how could non-kin altruism increase the frequency of the genes giving rise to it? The first thing to make clear is that Hamilton's theory does not *rule out* altruism among non-relatives. What it does do, however, is tell us that, unlike altruism among relatives, altruism among non-relatives cannot be selected *unless* there is some kind of return benefit to the

---

[12] Sherman (1977).

[13] See, e.g., Burnstein *et al.* (1994); Daly and Wilson (1988); Krupp *et al.* (2008); Stewart-Williams (2007, 2008).

[14] On vervet monkeys, see Seyfarth and Cheney (1984); on vampire bats, see Wilkinson (1984).

altruist. This insight opens the door to various other theories of the evolution of altruism. One of the most important of these is Robert Trivers' *reciprocal altruism theory*.[15] According to this theory, altruistic behaviour can enhance fitness as long as there's a sufficient probability that this behaviour will be reciprocated. One way to express the underlying logic of the theory is to say: you scratch my back and I'll scratch yours, and we'll both be better off than either of us would have been if we hadn't engaged in this mutual back-scratching.

There's a complication, though, and it's called *the problem of cheating*. Even when mutual exchanges of help are more advantageous than directly self-interested behaviour, it is more advantageous still to receive help but not return the favour. Individuals adopting this strategy are called *free riders*. Free riders will always do better than unconditional altruists, and thus if there were ever a population of unconditional altruists (an unlikely scenario), a mutant gene for free riding would spread like wildfire and ultimately displace the tendency to be altruistic. Is there any way around this problem?

There might be. One candidate approach would be to implement a 'tit-for-tat' strategy: help those who reciprocate your help, but stop helping those who do not. This would certainly minimize the losses due to free riders, but would it be successful enough to actually be selected? To answer this question, the political scientist Robert Axelrod set up a computer game in which, in each round, every player interacted with one other player, and had the option either to cooperate (i.e., help) or to defect (i.e., not help).[16] The players were not people, but computer programs. The programs embodied different strategies for playing the game, and were submitted by computer programmers from all around the world. In a series of tournaments, the various strategies were pitted against one another. There were simple strategies, such as 'Always Help' and 'Never Help',

---

[15] Trivers (1971).
[16] Axelrod (1984).

and also various more complex strategies. Then there was a strategy called 'Tit-for-Tat'. Tit-for-Tat always helped other players on its first encounter with them, and thereafter simply matched the other player's last move, helping if it helped but not if it didn't. The results took a lot of people by surprise. Tit-for-Tat was the most successful strategy by far. This is interesting, because Tit-for-Tat always came off worse whenever it first encountered a non-reciprocator: it would help but receive nothing in return. Nonetheless, it did best overall. The secret of its success was, first, that whenever it was short-changed in an interaction with a non-reciprocator, it saved itself further losses by immediately cutting off help to that individual, and, second, that it was able to profit greatly from its continuing and mutually beneficial interactions with fellow reciprocators.

Trivers proposed that certain emotional responses lead people to adopt a Tit-for-Tat-type strategy, without necessarily even being aware that that's what they're doing.[17] For example, the emotion of liking motivates us to initiate altruistic partnerships (Tit-for-Tat's 'cooperate on first move'). Anger and moral indignation motivate us to withdraw help to free riders or to punish them (Tit-for-Tat's 'defect if the other guy defects'). Gratitude and a sense of indebtedness motivate us to return good for good (Tit-for-Tat's 'cooperate if the other guy cooperates'). Guilt dissuades us from taking more than we give, thereby avoiding the possible negative consequences of being caught free riding; additionally, it motivates us to make reparations for free riding in the past. And forgiveness allows us to re-establish cooperative relations with reformed and repentant free riders. Trivers suggested that these emotional response patterns constitute an evolved moral psychology. In other words, many of our basic moral impulses were crafted by natural selection to facilitate reciprocally altruistic relationships and to deal with the problem of cheating.

[17] Trivers (1971, 1985).

Kin selection and reciprocal altruism are the best-known theories of the evolution of altruism, but there are others. According to one promising new approach, altruism has an evolutionary function comparable to that of the peacock's tail: although it doesn't increase the bearer's chances of surviving (if anything it does the opposite), it enhances fitness anyway because it is a costly display of an individual's genetic fitness, designed to impress mates or allies.[18] According to another theory, altruism is a product of group selection – that is, selection for heritable traits that advantage some groups over other groups, as opposed to advantaging some individuals over other individuals in the same group.[19] Finally, it has been argued that human altruism evolved through a process of cultural group selection, as opposed to genetic group selection: groups with the most adaptive cultural traditions survive for longer, and split off into new groups which 'inherit' the same traditions through the channels of cultural transmission.[20] Taken together, these theories provide a rich array of tools to explain altruistic behaviour. They constitute one of the great success stories of the evolutionary approach to psychology.

The accidental altruist

Nonetheless, the theories have attracted their fair share of criticism. Some theists doubt that they are up to the task of accounting for the full range of human altruism. Following C. S. Lewis, for example, the geneticist Francis Collins has argued that most of the acts we view as truly selfless are *not* explicable in evolutionary terms.[21] It is not uncommon for people to help individuals who are unrelated to them and who could never possibly reciprocate, and nor is it

---

[18] Miller (2000); Zahavi (1975).
[19] Darwin (1871); Fehr and Gaechter (2002); Sober and Wilson (1998); Wilson and Sober (1994).
[20] Boyd and Richerson (1985); Richerson and Boyd (2004).
[21] Lewis (1952); Collins (2006).

uncommon for people to help in situations that could not enhance their reputations or boost their value on the mating market. Evolutionary theory cannot explain this. Nor can it explain extreme cases of altruism. We've all heard stories about soldiers who throw themselves on live grenades to save unrelated comrades, for instance. Based on these kinds of considerations, Collins concludes that God, and not natural selection, must have gifted us our better natures.

As you've probably noticed, this is another God-of-the-Gaps argument, cast from the same mould as the argument from irreducible complexity discussed in Chapter 2, and the arguments discussed in Chapter 5 invoking God to explain the origin of life, the universe, and consciousness. As in those cases, even if there really were a gap here, this would not entitle us to assume that God fills that gap. But is there really a gap? Let's evaluate Collins' claims, starting with extreme altruism (e.g., the soldier who throws himself on a grenade). We should notice straight away that extreme cases are, by definition, rare. So even if we accept that extreme altruism is a gift from God, we'd have to admit that God has been rather stingy with his gift. If hurling oneself on top of grenades were the sort of behaviour we took for granted, the argument might have real force. However, the fact that such examples of self-sacrificial altruism grab our attention and command our admiration reminds us that this type of behaviour is *not* typical of our species. The type of behaviour that *is* typical – e.g., parents' altruism and selflessness towards their children – is the type of behaviour that evolutionary theory can readily explain. Thus, on closer inspection, the grenade example actually *supports* the idea that we are only partial altruists, and that the arsenal of theoretical resources provided by evolutionary theory goes a long way towards explaining standard levels of altruism in our species.

However, there still remains a stubborn core of unexpected altruism – altruism that is relatively common but that is not directed towards kin and not part of a network of reciprocal exchange. As I've already mentioned, such altruism may be explicable as a costly display of fitness, or in terms of group selection or culture (and I'll say more about the cultural component soon). But even if we ignore these possibilities, we still have no reason to invoke God. First, our unexpected altruism is not necessarily something that *needs* to be explained in evolutionary terms, at least not in the sense of trying to ferret out some evolutionary advantage to the altruist. In order for an evolutionary explanation to work, it is not the case that each and every instance of altruism needs to enhance the altruist's fitness. Natural selection sets in place behavioural dispositions that are advantageous *on average*. Non-adaptive altruism is not inconsistent with an evolutionary explanation, just so long as it's not the norm.

In addition, some examples of altruism may be evolutionary 'accidents': products of the mismatch between the environment in which our altruistic tendencies evolved and the strange modern environment in which we find ourselves today. For hundreds of thousands of years our ancestors lived in small communities, with a high density of kin and few strangers. The fact that we as a species are reasonably nice might be because the majority of our evolution took place in these conditions. In modern life, we are surrounded by large numbers of non-relatives and strangers, an evolutionarily novel circumstance. We may be a lot nicer than we would be if we were perfectly adapted to the sorts of environments we now inhabit. We may be able to preach, and sometimes even practise, an ethic of altruism towards all. If this, or some scenario like it, is correct, then our generalized altruistic tendencies are an evolutionary 'mistake'. Note that there is no implication that this is the kind of mistake we should seek to rectify. On the contrary, we can treasure this accidental pocket of non-adaptive kindness in this harsh Darwinian world. Unfortunately, it is possible that our generalized altruistic tendencies may be weeded out in the process of natural selection, if nice people in modern societies have fewer offspring on average than less nice people. As the proud owners of this accidental altruism, though, most of us probably hope that this will not happen.

Sexual morality

When it comes to moral behaviour, altruism has received the lion's share of evolutionists' attention. In fact, evolutionary psychologists sometimes act as if, in explaining altruism, they've explained the whole of morality. If pressed, however, they'll generally agree that altruism is merely one component of our moral repertoire, and that it's not the only one that invites an evolutionary explanation. Another is sexual morality. Given the importance of reproduction in our evolutionary history (and that of all sexually reproducing species), it is not surprising that many of the issues that fall within the compass of moral philosophy relate to reproduction. Examples include abortion, casual sex, contraception, homosexuality, infidelity, monogamy, polygamy, prostitution, and the sexual double standard. Some of our moral beliefs in these areas may be traced, directly or indirectly, to preferences and dispositions shaped by natural selection.

The area of sexual morality that most obviously lends itself to an evolutionary explanation is incest.[22] Incestuous mating is universally viewed as a moral no-no, and this makes good sense from a Darwinian perspective. Incest is a terrible move in evolutionary terms. Many recessive genes are detrimental to fitness if expressed. By definition, however, they are only ever expressed when both parents contribute the same recessive gene to their offspring. This is unlikely when the parents are unrelated. But close relatives have a greater than chance likelihood of sharing genes, and thus the offspring of incestuous liaisons are much more likely to end up with at least some paired recessives. In fact, it's virtually certain. The offspring of these liaisons have a reduced chance of surviving and reproducing, and thus, in the currency of reproductive fitness, they represent a poor investment of the parents' time and energy. For this reason, any psychological or

---

[22] Key works on this topic include Darwin (1871); Westermarck (1921); Wolf (1995).

behavioural tendency that decreases the likelihood of incestuous mating stands a good chance of being selected.

Consistent with this line of thought, it appears that human psychology includes a number of adaptations for incest avoidance. One such adaptation gives rise to what has been dubbed the *Westermarck effect*. This refers to the well-documented finding that people who grow up together tend not to be sexually attracted to one another.[23] Because humans usually grow up with kin, this tendency greatly decreases the chances of incestuous mating and its attendant dangers, i.e., damaged and deformed offspring. Some people believe that the dire consequences of incestuous mating are a punishment for the sin of incest. But this has things back to front. Incest does not have dire consequences because it's wrong; we consider it wrong because it has dire consequences. And this judgment of wrongness plausibly has an evolutionary origin. Admittedly, some incest does occur. However, this does not challenge the notion that we have adaptations for incest avoidance, any more than the fact that we sometimes end up as meals for other animals challenges the notion that we have adaptations for predator avoidance (e.g., the fight-or-flight response). Selection does not produce perfection. What it does, though, is help to explain some of our characteristic moral intuitions. Our moral judgments on the topic of incest, and on other aspects of sexual behaviour, can be traced back to emotions that have their origin in our evolutionary past.

Morality: easier said than done

We can see now how some of our basic moral inclinations and impulses make good sense in evolutionary terms. As I've tried to make clear, though, it's important to draw a distinction between our native moral psychology and our shared or formalized morality.

[23] See Bevc and Silverman (2000); Fox (1967); Lieberman *et al.* (2003, 2007); Shepher (1971, 1983); Westermarck (1921).

Shared morality does not have a direct evolutionary origin. It is a cultural construct built from raw materials furnished by natural selection, and it is the product of multiple forces. In this section, we'll consider some of the most important of these.

First, in some cases, the rules of morality are simply reflections of our evolved human nature. A good example relates to incest. Many cultures (though not all) have formal moral rules proscribing incest. It is tempting to ask why, if our aversion to incest has an evolutionary origin, we would need a moral injunction against it. But this question raises a possibility: maybe we *don't* need a moral injunction against incest. The vast majority of people just don't want to engage in this form of sexual expression; they feel an acute and pronounced disgust at the prospect. So, in this instance at least, formalized morality may simply embody or reflect people's values – values they would have anyway.

It's easy to multiply examples. We've seen that human beings are evolved nepotists, and it's clear that the ethical systems of the world reflect this aspect of human nature. The Old Testament, for example, admonishes us to honour our fathers and mothers,[24] and one of the key Confucian virtues is *filial piety*: love and respect for one's parents and ancestors. Likewise, humans are evolved reciprocators, and the reciprocity principle is embodied in many of our ethical teachings and spiritual beliefs. We like to believe that the scales of justice will be balanced, especially when this means that the good things we have done will be rewarded and the bad things other people have done will be punished. The reciprocity principle is embodied in the belief that God rewards and punishes behaviour, and also in the traditional Indian belief that one gets one's just deserts through the action of karma. There is also a widespread trend among human beings of

---

[24] Did you know that the recommended penalty for failing to honour one's parents is death? If you don't believe me, look it up! Try Exodus 21:15, 17; Leviticus 20:9; and Deuteronomy 21:18–21. Funnily enough, they don't tell you this kind of thing in Sunday School.

viewing disasters as a response to human misbehaviour, an exchange of bad for bad. The Chinese believed that earthquakes and other disasters occurred when people stopped following the way of nature. Similarly, the biblical flood was construed as a response to human sins. In these and other cases, we see that *universal aspects of the human mind shine through in the very different worldviews that arise in different cultures.*

There is clearly some truth, then, in the notion that our moral codes and ethical theories embody and reflect our evolved human nature. However, if pressed too far, this view becomes highly problematic. In some cases, morality redirects or even opposes our evolved passions. Although we have some degree of universal compassion, we also routinely favour ourselves, family members, and members of the groups to which we belong. As we've seen, this has a clear evolutionary origin. But most ethical systems frown on this kind of favouritism, which receives pejorative labels such as *selfishness, nepotism,* and *tribalism.* Similarly, although many moral precepts reflect the reciprocity principle (return good for good and bad for bad), this is not always the case. For example, the ethical precept that we should 'turn the other cheek' goes against the grain of our untrained inclinations. This demonstrates that our evolved moral psychology must be distinguished from our formalized morality, and that our official moral systems can and do stray away from the dictates of human nature.

Thus, we see that morality is more than just a reflection of evolved dispositions. This shouldn't come as too much of a surprise; if that were all there was to it, why would we spend so much time teaching our children what's right and wrong, and inculcating in them virtues such as generosity? And why, from time to time, would we all have such great difficulty following the dictates of morality? If morality were a direct and immediate product of evolution, it would presumably enhance our fitness and thus we would have evolved to want to be moral. Often, though, doing the right thing involves an effort of will and goes against our dominant impulse. As the philosopher Denis

Diderot pointed out, 'There is no moral precept that does not have something inconvenient about it'.[25] Certainly, in some domains, we *do* naturally want to do the right thing. Very few people have trouble resisting incest, for example. Nonetheless, we would often prefer to do the wrong thing if we thought we could get away with it. Thus some moral standards jar against our evolved inclinations. To explain this, we must go beyond biological evolution and consider the contribution of cultural evolution in the making of our moral systems.

Of course, when we have a conflict between a moral urge and a non-moral urge, this doesn't necessarily mean that the non-moral urge is an evolutionary product and the moral urge a product of culture. In some cases, it is a conflict between two evolved urges. Some evolved tendencies are considered moral; this includes the desire to care for one's children and to form monogamous pair bonds. But others are generally considered *im*moral; think, for example, of aggression, xenophobia, and our proneness to be unfaithful sexually.[26] Why do we consider some evolved motivations good and others not? One reason may be that familial love and monogamy do not impinge on the harmony of the community, whereas aggression and infidelity often do. This brings us back again to the idea that morality is not a direct product of evolution. Instead, and to some extent, it is a humanmade system of favouring those evolved tendencies that facilitate group cohesion, while disfavouring those that are socially divisive.[27] Morality is a way of controlling our evolved natures, rather than a mere reflection of those natures.

But morality isn't just for the good of the group. Other things are going on as well. First, some of the content of our moral systems derives from preferences about how we wish *others* to act. For example, people may tend to dislike selfishness in others and prefer

[25] Cited in Pratt (1994), p. 106.
[26] For good introductions to the evolutionary explanations for these behaviours, see Buss (2007); Pinker (1997); Ridley (1994); Wright (1994).
[27] Wilson *et al.* (2003).

altruism, especially when they or their kin are the recipients of that altruism. The moral codes of a culture may have emerged from people acting on such preferences – for example, discouraging others from being selfish and discouraging members of other families from exhibiting kin favouritism where this might disadvantage them (which is the point at which we stop seeing kin favouritism as wholesome family loyalty and start seeing it as nepotism). When you think about it, moral precepts such as 'give without thought of reward' are potentially highly advantageous to those who preach but don't practise them. Thus people may encourage others to behave altruistically in order to satisfy their own selfish agendas! Ironically, the self-interested, manipulative efforts of people pursuing this policy may have helped foster the unselfish forms of morality. This may help to explain why we are more altruistic than evolutionary principles alone would lead us to expect.

On top of that, our moral systems embody a desire to present ourselves in a good light. We may espouse ethical values that make us attractive as mates or friends or allies. This may mean understating the extent of one's bias towards self and kin. This isn't necessarily a matter of conscious and deliberate deception. Sometimes it is, but often we genuinely accept the moral principles we espouse, even when our behaviour doesn't quite match up to the rhetoric. The point is, though, that we sometimes display moral attitudes that convey a more flattering picture of ourselves than is strictly accurate, and these attitudes may embed themselves in the shared morality of a culture. Finally, we sometimes try to infer moral truths from the facts of our experience. For example, when any kind of disaster occurs, people may look at the events that preceded it and infer a causal connection. They may interpret it as a punishment for some moral misdemeanour, such as following the wrong religion or worshipping the wrong god or sacrificing the wrong animal in the wrong way. By finding spurious associations between our behaviour and subsequent events in the world, our moral codes come to be shaped in part by accidents of history. This introduces a random element into the moral code of any culture.

Ideas such as those above help to explain why our moral beliefs sometimes clash with our evolved predilections, and why they sometimes make little or no evolutionary sense. They are a crucial component of a complete explanation for human morality. Morality starts from evolved dispositions, but takes on a life of its own outside the individual's skull. Our shared moral code emerges over many generations as the net result of negotiations and deliberations and compromises among different people with different agendas. As such, when people learn the morality of the surrounding culture, they end up with something different between their ears than their evolved, morally relevant dispositions and inclinations.

Conclusion

In summary, there is reason to believe that certain facets of human morality have an evolutionary origin. However, our shared morality is not a product of evolution in any simple sense. It is hammered out in the process of social discourse. And although to some extent our moral codes embody and reflect human nature, to some extent they are also an *antidote* to human nature. Moral systems foster some evolved dispositions at the expense of others. Often they do this in the interests of the group. Sometimes, they do so in the interests of moralists – in other words, people promote certain moral beliefs because it is advantageous for them if *others* hold these beliefs. In short, morality is not a simple, unitary phenomenon that can be explained with any single theory or approach. It is a vast and messy conglomeration of phenomena, and must be understood as such.

There's one more thing to think about. The moral emotions – empathy, sympathy, and the like – evolved because they enhanced the fitness of their bearers, and for that reason their natural focus or range is relatively narrow. In the normal course of events, they are extended only to family and friends, and to members of our ingroup. But now that we've got these moral emotions, we can artificially extend them beyond their original domain. We can expand our moral circle to encompass strangers, members of other groups or nations, and even members of other species.[28] And we can do this not because it contributes to our fitness, but simply for the sake of making the world a nicer place, a place with less suffering. We can then sit back and ponder the fact that the cruel process of natural selection could give rise to organisms that are reasonably nice – strange monsters that recoil at the ruthlessness and cruelty of the process that created them, and that can strive to right the wrongs of nature.

---

[28] Singer (1981).

TWELVE

Remaking morality

The ultimate result of shielding men from the effects of folly, is to fill the world with fools.

Herbert Spencer (1868), p. 349

Just because natural selection created us doesn't mean we have to slavishly follow its peculiar agenda. (If anything, we might be tempted to spite it for all the ridiculous baggage it's saddled us with.)

Robert Wright (1994), p. 37

The nagging thought remains that Darwinism does have unsettling ethical consequences. The philosopher's reassurance that there will be no problem if we only remember to distinguish 'ought' from 'is' seems altogether too quick and easy. I believe this feeling of discomfort is justified.

James Rachels (1990), pp. 92–3

Moral guidance for the modern primate

In the last chapter, we saw that evolutionary theory sheds light on why we think that certain things are right and others wrong. But does it shed any light on what actually *is* right and wrong? A lot of people think that it does, or worry that it might. Since the nineteenth century, many thinkers have attempted to wrest moral principles from evolutionary theory. Some have concluded that the theory converges with Genesis in suggesting that we should be fruitful and multiply – in other words, that we should do whatever we can to survive and reproduce. Others have concluded that Darwin's theory supports a might-makes-right ethic, in which ruthlessness and selfishness are given the sanction of nature. The logic seems to be that natural selection is a vicious process, and a process that often favours the vicious, and thus that viciousness is somehow justified. And then there are the infamous *Social Darwinists*.[1] According to this motley collection of like-minded thinkers, evolutionary theory implies that society should be run according to the principle of the survival of the fittest, with unrestricted competition and no governmental aid for the poor or unsuccessful. Some have even argued that society's misfits should be prevented from reproducing, on the grounds that this would lower the genetic quality of the species. Ideas such as these were especially popular in the years and decades following the publication of the *Origin of Species*.

The question of the moral implications of evolutionary theory returned to centre stage in the 1970s with the rise of sociobiology and its close cousin, evolutionary psychology.[2] (Note that I'll generally treat these terms as synonymous.) Funnily enough, the best-known efforts to draw ethical implications from sociobiology come not from the sociobiologists themselves but from their critics. They therefore take the form 'sociobiology does not paint an accurate picture of human nature, but if it *did*, it would have all sorts of terrible ethical implications. Beware!' Note that the critics in this instance are not Creationists; they are 'Blank-Slate Darwinians'. These are individuals – well-intentioned, though somewhat outdated – who accept that we evolved but hold that the specifics of human behaviour are products of

---

[1] The classic work on Social Darwinism is Hofstadter (1944).
[2] See Segerstrale (2000).

culture, rather than manifestations of human nature. This applies particularly to things they don't like, such as selfishness, sexual possessiveness, and sex differences. Some Blank-Slate Darwinians claim that, if accurate, sociobiology and evolutionary psychology would support conservative right-wing political and social agendas, or even that these disciplines are pseudosciences specifically designed to buttress such ideologies. Some accuse advocates of these disciplines of trying to legitimate the status quo; excuse sexual infidelity and rape; and justify sexism, racism, and war. Some go further and credit evolutionary explanations of human behaviour with eugenics and the atrocities perpetrated by the Nazis. As members of the Sociobiology Study Group summed it up in a letter to *The New York Review of Books*:

> The reason for the survival of these recurrent determinist theories is that they consistently tend to provide a genetic justification of the *status quo* and of existing privileges for certain groups according to class, race, or sex … These theories provided an important basis for the enactment of sterilization laws and restrictive immigration laws by the United States between 1910 and 1930 and also for the eugenics policies which led to the establishment of gas chambers in Nazi Germany.[3]

It's important to point out that the sociobiologists and evolutionary psychologists themselves have never claimed that their work implies anything like this. In fact, they have rarely claimed that it has any ethical implications at all. A notable exception is E. O. Wilson, who, in his voluminous writings on sociobiology, has made periodic forays into moral philosophy. This has made him a highly controversial figure. However, most of the ethical conclusions he has drawn have been much milder and less controversial than those attributed to him by his panicky critics. Wilson has argued, for instance, that sociobiology provides a rationale for universal human rights, the moral permissibility of homosexuality, and the preservation of the human gene pool, and

[3] Sociobiology Study Group for the People (1975), p. 43.

biodiversity on this planet.[4] These are all claims unlikely to raise most critics' hackles.[5] Nonetheless, it is the implications drawn by the critics that are linked most closely with sociobiology in most people's minds: sociobiology justifies the status quo; lets men off the hook for their sexual transgressions and betrayals; and implies that cut-throat capitalism is the only workable political system.

For those who wish to reject these implications, or any of the implications we've been discussing, several options are available. One is to reject the truth of evolutionary theory. A more reasonable alternative is to argue that evolutionary theory does not have the ethical implications that are claimed. And one way to do *that* is to argue that evolutionary theory has no ethical implications whatsoever. This is a popular position, and has doubtless been motivated in part by some of the unpopular and unpleasant conclusions that have been drawn from Darwin's theory in the past. But even if that's the main motivation, the argument most commonly offered to support the no-implications position is the idea that evolution deals with facts rather than values, and that values cannot be derived from facts. Over the last few decades, this has emerged as a virtual consensus, championed by some of the biggest names in biology, including Stephen Jay Gould and Richard Lewontin. There is, however, a rising tide of dissenters and an important middle ground to explore. That's the aim of this chapter. Our conclusion will be that evolutionary theory does not have any of the implications typically attributed to it, but that this is *not* because evolutionary theory has no ethical implications at all.

Hume's guillotine

Does the fact that a behaviour evolved imply that it is morally permissible? A lot of people assume it does, and it's easy to see why this

---

[4] Wilson (1978, 1984, 1999).

[5] For other claimed implications of evolutionary theory for ethics, see Arnhart (1998); Hursthouse (1999); Railton (1986); Richards (1986); Rosenberg (2000, 2003); Singer (2000).

assumption seems reasonable – even tempting – when first encountered. We don't think that lions are immoral for killing gazelles, even if we find it gruesome; we wouldn't lock them up for transgressing the gazelles' right to life and nor would we take action to prevent them from doing it again. It's just the way nature is, we think. Why is it different when it comes to human beings? This is not intended as a rhetorical question, and in fact many attempts have been made to answer it. The most common answer is that ethical systems derived from evolutionary theory commit an error of reasoning known as the *naturalistic fallacy*. When people invoke this term, they're usually thinking of the fallacy of assuming that because something is natural, it must necessarily be good, or that the way things are is the way they ought to be. Technically, the naturalistic fallacy is neither of these things. The term was introduced by the philosopher G. E. Moore, and refers to something far less intuitively obvious.[6] The fallacy outlined by Moore has not fared as well as the term he coined for it, so we'll limit the discussion to the more usual understanding of the naturalistic fallacy.

Our starting point is our good friend David Hume. More than a century before Darwin unveiled his theory, Hume pointed out a recurring error in discussions of ethics:

> In every system of morality, which I have hitherto met with, I have always remarked that the author proceeds for some time in the ordinary way of reasoning ... when of a sudden I am surprised to find, that instead of the usual copulations of propositions, is, and is not, I meet with no proposition that is not connected with an ought, or an ought not. This change is imperceptible; but is, however, of the last consequence. For as this ought, or ought not, expressed some new relation or affirmation, 'tis necessary that it should be observed and explained; and at the same time that a reason should be given, for what seems altogether inconceivable, how this new relation can be a deduction from others, which are entirely different from it.[7]

---

[6] Moore (1903).

[7] Hume (1739), Book III, Section I, Part I.

In other words, when arguing that something is morally right or morally wrong, people often begin by making factual assertions ('is-statements'), but then somewhere along the line shift quietly to making evaluative assertions ('ought-statements'). The leap from factual premises to evaluative conclusions is not deductively valid. Consider the following Social Darwinist-style argument:

> Efforts to aid the weak, sick, or poor go against nature.
> *Therefore*, we ought not to aid the weak, sick, or poor.

The premise of this syllogism (the first line) does not entail the conclusion, because the conclusion contains an element not present in the premise: the evaluative element, represented in this case by the word *ought*. Thus, even if the premise were true, the argument would not be valid and would fail to establish its conclusion. As Hume noted, no collection of solely factual premises could entail any evaluative or moral conclusion. This principle is sometimes known as *Hume's law* or *Hume's guillotine*.[8] If it were truly appreciated, Hume suggested, it would 'subvert all the vulgar systems of morality'.

Hume's law is usually viewed as a knock-out argument against evolutionary ethics in its varying forms. It seems plainly obvious that you can't derive an 'ought' from an 'is'. It seems so obvious, in fact, that maybe we should be worrying that we're missing something. Hume's law seems to show that facts about evolution can have no bearing on ethical issues, and that factual and ethical reasoning are completely independent domains of discourse. But it does not have this implication at all. The importance of the is-ought fallacy has been drastically overstated. Consider this argument again:

> Efforts to aid the weak, sick, or poor go against nature.
> *Therefore*, we ought not to aid the weak, sick, or poor.

---

[8] Hume is one of those lucky people who have had various ideas named after them. The most famous is *Hume's fork*.

Clearly, the argument is not deductively valid. This could easily be remedied, however, by including an additional premise that would justify the leap from *is* to *ought*. After all, it is possible in principle to construct a valid argument from *any* premise to *any* conclusion, given the appropriate intervening premise.

> Efforts to aid the weak, sick, or poor go against nature.
> We ought not to go against nature.
> *Therefore*, we ought not to aid the weak, sick, or poor.

The argument is now deductively valid, and thus if the premises are true, the conclusion must be true also. Furthermore, the new premise is one that many would accept. The idea that we ought not to go against nature is commonly heard in arguments against genetic engineering and other new biotechnologies. Let me make clear that I'm not suggesting that we ought *not* to aid the weak, sick, or poor. My point is simply that if there's anything wrong with the above argument, it's not that it's deductively invalid. It does not commit any *logical* fallacy. Soon I'll argue that arguments such as these rest on false assumptions about evolutionary theory. Our concern for the moment, though, is the more general point that Hume's law does not rule out the possibility that factual premises can inform evaluative conclusions, as long as the former are conjoined with premises that are themselves evaluative.

A useful distinction can be drawn between *ultimate ethical statements* and *derived ethical statements*. Ultimate ethical statements are our foundational ethical commitments; they are not implications of other, more general ethical statements. Derived ethical statements are statements that are deduced from ultimate ethical statements taken in conjunction with factual statements. In the above example, the ultimate ethical statement is 'We ought not to go against nature', and the derived ethical statement is 'We ought not to aid the weak, sick, or poor.' Although we cannot read ultimate ethical values straight from the facts of evolution, this does not rule out the possibility that such facts could figure into our moral reasoning. By combining ultimate moral principles with facts about the world gleaned from evolutionary biology, we may be able to generate new and important moral conclusions.

Of course, we then come face-to-face with a new problem: where do we get our ultimate moral principles from? Do we just make them up? If so, then rather than accepting ultimate values based on facts about evolution, we accept ultimate values based on – absolutely nothing! I will postpone discussion of this thorny issue till Chapter 14. For now the key point is that there is no logical barrier preventing us from deriving ethical conclusions from facts about the world. Having established this point, we can next ask what ethical implications evolutionary theory might have.

Does evolutionary psychology justify the status quo?

To begin with, let's consider some of the putative implications suggested by the critics of sociobiology and evolutionary psychology. We'll start with the most common suggestion of them all: that evolutionary explanations of behaviour serve to justify the status quo. One critic, Dorothy Nelkin, expressed this viewpoint when she asserted that evolutionary psychology is 'a way to justify existing social categories', and that one of its implications is that no 'possible social system, educational or nurturing plan can change the status quo'.[9] This supposedly applies, among other things, to gender roles. Sociobiologists and evolutionary psychologists claim that, on average, women have a stronger parental motivation than men, and that this has its origin in our evolutionary history.[10] Some early critics interpreted this as the

[9] Nelkin (2000), p. 22.

[10] The claim that women have a stronger average parental urge than men is sometimes viewed as a sexist generalization. But it's only sexist if we take a dim view of the trait in question: the parental urge. One could turn the accusation on its head: those who view the evolutionist's claim (that women are more parental than men) as sexist are actually being sexist themselves, because they're taking a negative view of a trait that's usually found more strongly in females than males. They are therefore prizing prototypically masculine traits more highly than prototypically feminine ones.

claim that women should be kept barefoot and pregnant, and be denied access to the workplace. Obviously, they weren't too pleased by this. Other critics reached equally dramatic conclusions on other topics. For example, some claimed that sociobiology, if true, would justify existing inequalities between different classes and races. More generally, the concern was that evolutionary explanations of human behaviour imply that the way things are is the way things ought to be.

I should emphasize again that such implications have never been claimed by sociobiologists or evolutionary psychologists themselves. Indeed, most of them emphatically disavow such implications. Often they cite the is-ought fallacy as their grounds for doing so. As we've just seen, though, the is-ought fallacy does not rule out the possibility of an evolutionarily informed argument supporting the status quo. If we hope to reject these views, we must look for other reasons. And of course if sociobiology implies that we *should* preserve the status quo, and if sociobiology turns out to be an accurate science, then we should preserve the status quo. Right? But let's see if it really has this implication. Consider the following argument:

> The status quo has its origin in our evolutionary history.
> *Therefore*, we should maintain the status quo.

This argument jumps straight from an is-statement to an ought-statement, and is therefore not deductively valid. This is easily rectified, however. One way to patch up the argument would be as follows:

> The status quo has its origin in our evolutionary history.
> We should maintain anything that has its origin in our evolutionary history.
> *Therefore*, we should maintain the status quo.

So there we have it: a deductively valid argument for maintaining the status quo! But is the argument sound? (In logic, an argument is sound when it is deductively valid *and* its premises are true.) The first thing to notice is that precisely the same argument could be used just as effectively with a *non*-evolutionary explanation for the status quo.

> The status quo has its origin in cultural processes.
> We should maintain anything that has its origin in cultural processes.
> *Therefore*, we should maintain the status quo.

The obvious retort is: why should we assume that something should be maintained just because it has its origin in cultural processes? Why indeed! However, we have precisely as much reason to ask why we should assume that something should be maintained just because it has its origin in our evolutionary history. This is really the crux of the argument against deriving morals from evolution: there is no good justification for it. We can do it if we want to, I suppose, but why would we want to? Even if a given aspect of the status quo has its origin in our evolutionary history, we have no moral obligation to preserve it. Thus, the fact that, on average, women are more parental than men does not imply that women *should* be more parental than men (or that they should not), and nor does it imply that any particular female has to be a parent if she does not wish to. Evolutionary psychology tells us why there is an average sex difference in this domain, but it provides no reason to think that individuals should model themselves on the average. It offers explanations, not ethical advice or policy recommendations, and we're free to try to change things if we so desire. That said, one might ask whether, if something has an evolutionary origin, it is actually possible to change it. And in some cases it may indeed be difficult. But if there is *any* chance of changing things, presumably this will not be achieved by insisting that we affirm the view that everything is down to culture and that evolution has nothing to do with it. Our best bet is to make an honest analysis of the causes of our circumstances and then let this guide our efforts to change things, if we decide that that's what we want to do.

Does evolutionary psychology eliminate responsibility?

Another common complaint rolling off the tongues and out of the pens of the critics of sociobiology is based on the idea that evolutionary explanations for objectionable behaviour have the effect of excusing that behaviour. Evolutionary psychologists argue that various kinds of antisocial behaviour, including aggression, infidelity, and perhaps even rape,[11] can be traced to evolved motivations and emotional dispositions. Surely, though, if evolution 'made us do it', we cannot take the blame; we become mere innocent bystanders of our own behaviour. This, at any rate, is a commonly held view. But is it true? Do evolutionary explanations for undesirable behaviour imply that people can never be held accountable for their moral transgressions? Or worse, have these explanations been deliberately engineered by men to let men off the hook?

To address this issue, we'll focus on one claim in particular, an oldie but a goodie. It comes from the biologist Steven Rose. Discussing sex differences in sexual behaviour, Rose wrote that, according to socio-biologists, 'human males have a genetic tendency towards polygyny, females towards constancy (don't blame your mates for sleeping around, ladies, it's not their fault they are genetically programmed)'.[12] Rose's comment was clearly intended as a criticism of sociobiology, although it's not clear exactly what the criticism is – that sociobiology has unfortunate implications and therefore cannot be true? That sociobiologists (the male ones at least) are deliberately concocting an explanation that will let them get away with sleeping around? Neither criticism is worth taking seriously, but there's an assumption underlying both that we need to evaluate, namely, that if men's tendency to sleep around and stray sexually has an evolutionary origin, men cannot be held accountable for this behaviour.

Let's start by spelling out the obvious: even if a genetic tendency towards polygyny really did imply that we couldn't blame men for

---

[11] Thornhill and Palmer (2000).
[12] Rose (1978).

sleeping around, it could still be the case that men have a genetic tendency towards polygyny. There is no law of nature guaranteeing that the universe will be such as to make members of one particular primate species happy. The next thing to point out is that evolutionary psychologists *don't* argue that men have a genetic tendency towards polygyny and women towards constancy (i.e., monogamy). We're a relatively monogamous species overall – women and men – and this is the norm for those rare species in which members of both sexes invest heavily in offspring. (In most mammals, it's only the females.) And although we do sometimes cheat, men are not the only guilty parties in this respect. Women do too on occasion, and there are various theories about how this could have evolved.[13] What evolutionary psychologists actually claim is this: *on average*, men have a stronger desire than women for sexual novelty and casual sex, and *on average*, men are less choosy about sexual partners in a short-term mating context – i.e., for a one-night stand or brief affair. Men are about as choosy as women when it comes to long-term or marital partners.[14]

Next we should examine the general assumption underlying Rose's claim: that if human behaviour can be explained in evolutionary terms, this implies that we cannot be blamed for our bad behaviour (or, presumably, praised for our good behaviour). Let's start by accepting, for the sake of argument, that this is true. The first thing we have to ask is whether non-evolutionary explanations have a *different* implication. It is surreptitiously assumed that they do, but we should question the

---

[13] Strangely, though, evolutionary psychologists – even the many female ones – are never accused of trying to excuse female infidelity. It seems to work like this: when the conclusions of evolutionary psychologists are unflattering to females, it is sexist. But when the conclusions are unflattering to males – for example, when it is claimed that males are predisposed to philander, act violently, or commit rape – this is giving men permission to engage in these activities. It's a double standard, but with females on the advantaged end.

[14] See, e.g., Buss (1994); Buss and Schmitt (1993); Clark and Hatfield (1989); Diamond (1997b); Ellis and Symons (1990); Kenrick *et al.* (1990); Schmitt (2005); Symons (1979).

assumption. Most opponents of evolutionary psychology concede that men have a stronger desire for casual sex than women; they simply provide an alternative, sociocultural explanation for the inclination (for example, they suggest that men are taught by their culture to prize sexual conquest). Why not argue that the sociocultural origin of the tendency to cheat implies that men cannot be held accountable for cheating? If an evolutionary explanation exonerates a person, surely a non-evolutionary explanation does exactly the same thing. Rose could just as well have said: 'Don't blame your mates for sleeping around, ladies, it's not their fault they were born into a patriarchal society!'

It might be argued, though, that there's an important difference between the evolutionary and the cultural explanations. The evolutionary explanation implies that the behaviour in question is 'in our genes'; we don't choose our genes and thus we cannot be held responsible for what our genes have wrought. Even if we were to accept this argument, however, the blank-slate position is no better. We don't choose the culture into which we're born or our early-life environments, any more than we choose our genes. So, if we can't be held responsible for the effects of our genes, how could we be held responsible for the effects of these other things? Admittedly, we choose some of our later environments. But we don't choose the genetic or environmental influences that determine the choices we make. And although we actively shape the world around us, the ways in which we do so are themselves causally determined from moment to moment. Ultimately, who we are and what we do can be traced back to causes outside ourselves, in principle if not in practice, and this is the case whether we think that evolution has crafted specific psychological and behavioural tendencies (including sex differences in sexual inclinations), or we think that evolution has made us blank slates and that any sex differences are learned. Either way, we are not the *ultimate* causes of our own decisions or actions.[15]

---

[15] Strawson (1986). For an excellent introduction to this issue, see Radcliffe Richards (2000), chapter 6.

The upshot is that, if an evolutionary explanation eliminates responsibility, so does a sociocultural one. But does an evolutionary explanation eliminate responsibility? No! There's no reason at all to think that it does. To understand why not, we need to think about the purpose or function of the concept of responsibility. What's it for? In essence, the function of holding people responsible for their actions – the ultimate reason we do it – is to improve their behaviour and protect society. Why should these goals be acceptable only if people's behaviour is inexplicable or uncaused? Why should we only try to prevent undesirable behaviour that cannot be traced to natural selection? It makes no sense. Surely we can seek to improve behaviour and safeguard society regardless of whether people's behaviour can be traced to causes outside themselves, and regardless of whether these causes are genetic or cultural or – as is almost always the case – a combination of the two. We have no reason to think that evolutionary explanations eliminate responsibility.

Does evolutionary theory lead to Social Darwinism?

The next putative implication to consider is the idea that evolutionary theory or sociobiology leads us inexorably down the dark path to Social Darwinism. As I mentioned earlier, the Social Darwinists argued that we should apply (supposed) Darwinian principles to society. This was done to justify everything from war and imperialism to racism and patriarchy.[16] But by far the best-known application of Darwinian

---

[16] Paul (2003). Note that evolutionary theory has also been used to justify a very different political agenda. Thus, whereas some used the theory to justify war (on the grounds that it was the way of nature and was ultimately beneficial for the species or civilization), others used it to argue for peace and the rejection of militarism (on the grounds that many fit young men were killed during war and thus prevented from breeding, while the physically inferior remained at home with the women) (Jordan, 1915). Similarly, whereas some used Darwin's theory to justify patriarchy (on the grounds that men had supposedly evolved to be physically and mentally superior to women), others found support for feminism (on the

principles was to the marketplace.[17] The Social Darwinists held that society should be run in accordance with the principle of the survival of the fittest, and thus advocated dog-eat-dog capitalism and an extreme laissez-faire approach to government (i.e., little or no state intervention). They attempted to ground their recommendations in facts about evolution. Although it wasn't always spelt out, the rationale for a sink-or-swim society went something like this. Conflict and competition between individuals, groups, and nations produce progress. Therefore, if society is set up so that everyone competes against everyone else, the best competitors will prevail and there will be ongoing societal improvement. Because natural selection is inherently progressive (or so it was claimed), a sink-or-swim society would automatically and inevitably produce progress. The engine of evolutionary change is the struggle for existence. To remove this struggle would quash progress and foster degeneration. We therefore have a moral obligation not to interfere with the course of natural selection, or even have an active obligation to help the evolutionary process on its merry way.

Many Social Darwinists opposed governmental interventions to help the weak, sick, and poor, on the grounds that these efforts were self-defeating in the longer term. As the philosopher and self-educated maverick Herbert Spencer put it:

> To aid the bad in multiplying, is, in effect, the same as maliciously providing for our descendants a multitude of enemies. Doubtless, individual altruism was all very well, but organized charity was intolerable: unquestionable injury is done by agencies which undertake in a wholesale way to foster good-for-nothings; putting a stop to that natural process of elimination by which society continually purifies itself.[18]

grounds that the subjugation of women undermined female sexual choice and was therefore a threat to the quality of the species). None of these arguments are particularly compelling; my point is simply that evolutionary theory has been used to justify a wide range of mutually inconsistent ideologies, and that this casts doubt on the idea that it actually justifies any.

[17] See, e.g., Spencer (1874); Sumner (1883).   [18] Spencer (1874), p. 345.

Spencer also pointed out that, when aid comes from the government, taxpayers have no choice but to shoulder the burden of providing for their descendants 'a multitude of enemies'. (Strictly speaking, they do have a choice: shoulder the burden or go to prison.) In his opinion, unfettered competition would eventually eliminate these unfortunate blights upon humanity. We are evolving towards moral perfection, but unfortunately the price for perfection is the suffering of 'good-for-nothings'. Their hardships and misfortunes are the growing pains of civilization. Our attempts to intervene in the natural scheme of things, however well motivated they might be, will forestall the process and ultimately make matters worse.

Many of America's leading businessmen and industrialists found moral support for an unrestrained free market in Darwin's theory and Spencer's writing. This included John D. Rockefeller and Andrew Carnegie. Rockefeller apparently informed a Sunday School class that 'the growth of a large business is merely a survival of the fittest ... This is not an evil tendency in business. It is merely the working out of a law of Nature and a law of God.'[19] In the same spirit, Carnegie suggested that we accept and welcome 'great inequality of environment, the concentration of business, industrial and commercial, in the hands of a few, and the law of competition between these, as being not only beneficial, but essential to the future progress of the race [i.e., species]'.[20] (Coincidentally, the things that Carnegie viewed as beneficial to the species also happened to be beneficial to him.)

To be fair, some of the more eloquent Social Darwinists (Spencer in particular) did more than just use Darwin's theory to justify harsh social and economic practices. But it is those conclusions that have tarnished evolutionary theory by association, and thus it is those conclusions we must wrestle with. On the basis of Hume's law, we can reject any Social Darwinist argument that proceeds directly from *is-statements* to *ought-statements*. However, this only rules out a certain

---

[19] Cited in Hofstadter (1944), p. 45.
[20] Carnegie (2006), p. 3.

class of *arguments*. It does not show the falsity of Social Darwinist *conclusions*. So let's consider on what grounds it might be argued that society should be organized according to the principle of the survival of the fittest. One approach would be to argue that it is the way of nature and that the way of nature is good. This treats the survival of the fittest as *a good thing in itself*. An alternative would be to argue that the way of nature is *a means to other ends*. As we've seen, the Social Darwinists were impressed by the idea that evolution involves progress, and thought that the crucial ingredient producing this progress was the survival of the fittest: the struggle of individual against individual, group against group. This provides the means to ends that, quite independently of evolution, we consider good. State interference and social welfare are undesirable, not because they go against the way of nature per se, but because the way of nature produces progress.

We may wish to ask whether the means justify the ends. However, there are more fundamental problems with this whole line of thought. The Social Darwinists misunderstood the evolutionary process in a number of crucial ways. One is the idea that to oppose the survival of the fittest – i.e., to be cooperative and altruistic – is to go against nature. The phrase 'survival of the fittest' is somewhat misleading. (The phrase is Spencer's, incidentally, not Darwin's, although Darwin did eventually adopt it.) As we saw in the last chapter, evolutionary history is not simply a Hobbesian war of all against all. There is plenty of warring and competitiveness in nature, certainly, but natural selection can also produce cooperation and altruism among organisms. Thus such tendencies do not necessarily go against nature. Natural selection is ruthless, but not all its products are.

According to most evolutionary biologists, it is only at the level of the gene that nature operates according to principles analogous to those that the Social Darwinist favours. Genes are 'selfish', in the sense that the genes that persist in the population are the ones that have phenotypic effects that benefit them – i.e., that result in them being copied at a greater rate than competing versions of the same genes.[21] Certainly, different genes often 'cooperate' with one another (for example, to build coherent organisms that preserve and propagate them). However, they do so only if it is in their own interests. They are 'selfish cooperators', to use Dawkins' terminology.[22] There is no altruism among genes, and no genetic equivalent of social welfare. In order to salvage his position, the Social Darwinist would have to maintain the (rather peculiar) view that human societies ought to mimic the conditions under which gene selection takes place. But to establish such a social system, we would have to artificially suppress our natural altruistic tendencies, and artificially emphasize our natural selfish and competitive tendencies. Consequently, to pursue this line of argument, the Social Darwinist would have to let go of the idea that we should follow the way of nature.

That would leave the argument that natural selection produces progress and all things good. As we saw in Chapter 9, though, the idea that evolution is progressive is deeply flawed. Let me recap the argument. First, evolution is not a matter of ongoing betterment. As the environment changes, the criteria for goodness of design change with it. More importantly, natural selection favours any trait that increases the likelihood that the genes giving rise to it will be copied, regardless of whether we consider the trait in question to be good or desirable in any way. What's adaptive in evolutionary terms is not always what's desirable to us. Although selection accounts for some things we view as good (altruism, for example), it also accounts for plenty we consider bad. As a result, there is no reason to think that the selective principles that operate among genes must necessarily lead to the betterment of society or to an increase in the sum total of human happiness. There is therefore no reason to think that society should be run according to these principles. This undermines the Social Darwinist viewpoint, and

---

[21] On the selfish gene theory, or genes'-eye view of evolution, see Dawkins (1982, 1989); Hamilton (1963, 1964); Williams (1966).
[22] Dawkins (1998b).

shows that there is no necessary connection between evolutionary theory and laissez-faire social policies. If anything, the suffering produced by natural selection could support an argument *against* such policies.

Does evolutionary theory lead to eugenics?

Closely related to Social Darwinism is the hot-button topic of eugenics. Eugenics can be defined as any deliberate effort to improve the genetic constitution of the human species, or to prevent it from deteriorating. This usually involves the application of the methods of selective animal breeding to human beings. Plato was the first to describe how this might work, and his description was rather blunt: 'The best men must have sex with the best women as frequently as possible, while the opposite is true of the most inferior men and women.'[23] The founder of modern eugenics (and the person who coined the term *eugenics*) was Darwin's cousin, Francis Galton. Galton was deeply concerned that the genetic quality of the species was deteriorating. He attributed this to two main factors. The first was that, in modern technological societies, the cleansing hand of natural selection had been removed, and thus 'unfit' specimens of humanity were now able to survive and reproduce. Galton started worrying about this possibility after reading the *Origin*; later, Darwin read Galton's work and started worrying about it himself. In *The Descent of Man*, he wrote:

> With savages, the weak in body or mind are soon eliminated ... We civilised men, on the other hand, do our utmost to check the process of elimination; we build asylums for the imbecile, the maimed, and the sick; we institute poor laws; and our medical men exert their utmost skill to save the life of every one to the last moment ... Thus the weak members of civilised society propagate their kind. No one who has attended to the breeding of domestic animals will doubt that this must be highly injurious to the race of man ... excepting in the

[23] Plato (1992), 459D.

case of man himself, hardly any one is so ignorant as to allow his worst animals to breed.[24]

Another reason for the genetic decline of the species, according to Galton, was that, in modern societies, the people who have the most offspring are not the most intelligent, the most industrious, or the most conscientious. Instead, the unintelligent have more offspring than the intelligent, the lazy more than the industrious, and the reckless more than the conscientious. (Proving his own point, Galton himself remained childless.) Left unchecked, suggested Galton, this phenomenon (now known as *dysgenic fertility*) would result in the degeneration of the species and the collapse of civilization.[25] One of the main purposes of eugenics was to prevent this from happening.

Eugenics became more and more popular (and less and less scientifically rigorous) throughout the first half of the twentieth century, and many countries adopted eugenics policies. This included the compulsory sterilization of those deemed criminal, mentally retarded, or mentally ill, and the institution of marriage laws and strict immigration policies. Before Hitler, the most extensive eugenics programme by far was found in the United States. But of course it was the Nazis who took the trend to its extreme, sterilizing or murdering thousands upon thousands of 'undesirables', and attempting to wipe out the European Jews. When the nightmarish excesses of the Nazi programme were revealed after World War II, there was a massive backlash against eugenics. The topic is now an intellectual taboo.

So here's our question: does the truth of evolutionary theory imply that we must accept the necessity of eugenics? Many of the most important evolutionary biologists of the first half of the twentieth century were supporters of eugenics. Virtually none of them are now.

---

[24] Darwin (1871), p. 159. Note that, despite his concerns, Darwin did not think that we should stop caring for the sick or the poor. To do so would be to lose touch with the best part of our humanity.

[25] See Lynn (1996) for a more recent version of this argument.

This shows (among other things) that acceptance of evolution does not *compel* the acceptance of eugenics. Conversely, eugenics could easily have proceeded even in the absence of evolutionary theory, just as the selective breeding of animal and plant stock did for thousands of years. At the most, Darwin's theory raises the salience of the relevant issues and starts people thinking about the possibility that eugenics policies might work.[26] It says nothing about whether we *should* implement such policies. The answer to that question cannot be read directly from nature or from the laws of evolution. It is something we have to decide for ourselves.

Although most people unambiguonsly oppose eugenics, a small number of scholars think we should think again. The psychologist Richard Lynn argues that the condemnation of eugenics in the latter half of the twentieth century went too far.[27] The Nazis gave eugenics a bad name (to put it mildly), leading the public to conflate it with genocide and involuntary euthanasia. But there are versions of eugenics that might be consonant with modern sensibilities. In particular, suggests Lynn, modern biotechnologies, such as genetic screening and genetic engineering, may open up the possibility of a voluntary and undirected form of eugenics. In taking advantage of these technologies, people will exert greater choice over the constitution of their offspring; en masse, they may direct the constitution of the gene pool. The key point, though, is that if we decide to permit humane forms of eugenics, or if we decide to ban them, this is our decision. Evolutionary theory does not license or mandate either option.

---

[26] Critics often claim that eugenics would not and could not work. But eugenics has been employed since the dawn of agriculture: people have selectively bred animals and plants possessing desired traits, and have done so with a great deal of success. There is no reason to suppose that it wouldn't work with humans as well. It would probably be *politically* impossible to institute an effective eugenics programme in the West given today's intellectual climate, but it surely isn't biologically impossible (Lynn, 2001).

[27] Lynn (2001).

Did evolutionary theory lead to Nazism?

We've already touched on the supposed link between Darwinism and Nazism, but it's an important issue and we should examine it in more detail. Over the last half century, many Creationists and Intelligent Design theorists have attempted to blame the Nazi atrocities on evolutionary theory. For example, according to ID advocate Richard Weikart, author of the book *From Darwin to Hitler*, evolutionary theory directly contributed to the rise of the Nazis and their attempt to exterminate the Jews:

> Darwinism by itself did not produce the Holocaust, but without Darwinism, especially in its social Darwinist and eugenics permutations, neither Hitler nor his Nazi followers would have had the necessary scientific underpinnings to convince themselves and their collaborators that one of the world's greatest atrocities was really morally praiseworthy. Darwinism – or at least some naturalistic interpretation of Darwinism – succeeded in turning morality on its head.[28]

Admittedly, if you read some of what Hitler and his cronies said, it certainly had the trappings of evolutionary science. For instance, Hitler once claimed that:

> If we did not respect the law of nature, imposing our will by the right of the stronger, a day would come when the wild animals would again devour us ... By means of the struggle the elites are continually renewed. The law of selection justifies this incessant struggle by allowing the survival of the fittest.[29]

Is it fair, though, to pin Nazism and the Holocaust on evolutionary theory? A common response to the charge that Darwinism led to Nazism – and I think a very reasonable one – is that Darwinian ideas and phrases were simply attached to streams of thought that existed already anyway. The ideological roots of Hitler's philosophy (e.g., the

---

[28] Weikart (2004), p. 233.
[29] Cited in Trevor-Roper (1963), pp. 39, 51.

notion that the Germans were the supreme race and that they were under threat from the Jews) existed long before evolutionary theory arrived on the scene. The theory was used to buttress existing attitudes, rather than giving birth to those attitudes in the first place. Terms and phrases such as 'struggle' and 'competition for survival' served useful rhetorical purposes, and the link to evolutionary theory – however tenuous – lent Nazism scientific credibility, at least among the credulous. But although the Nazis used terms and phrases culled from Darwinian thought, their brand of racial science was pseudoscience. Their theories were based on racist attitudes, not on evidence, and they did not critically examine the theories or expose them to empirical scrutiny. Most modern evolutionists deny that natural selection has produced any significant differences between human racial groups, and such differences as do exist are due mainly to differences in the environment as opposed to differences in the frequency of particular genes across human populations.[30] Scientifically and intellectually, the Nazis didn't have a leg to stand on.

It might still be argued that Darwinism fuelled the fires of Nazi philosophy, enhancing the perceived authority of the pre-existing streams of thought. But what should we conclude from this? That because a bastardized version of evolutionary theory was used to help justify something terrible, the theory isn't true? That although evolutionary theory is true, we should suppress knowledge of it? No. Like Social Darwinism, Nazi philosophy was riddled with misunderstandings about evolution. Rather than providing a reason not to teach

[30] Even if the differences *are* due in part to genes, there would be two things to remember. First, we'd be talking about relatively trivial *average* differences between groups, not differences that distinguish all or most members of one group from all or most members of any other. And, second, we should always evaluate individuals on the basis of the actual traits they possess, not on the basis of simple statistical descriptions of the groups to which they belong. If someone has an IQ of 100, for example, what does it matter whether the mean IQ for that individual's racial group is higher or lower than that?

evolutionary theory, Nazism provides a reason to teach it as clearly and as carefully as possible.[31]

There's another point to make. If one wishes to put some of the blame for Hitler on Darwin, then to be consistent one must also put some of the blame on Christianity. Although not a practising Christian, Hitler was quite possibly a theist. He often claimed that, in trying to rid the world of the Jews, he was doing God's work. And even if he didn't really believe this, many of his followers did. Indeed, *most* of those who carried out Hitler's orders were Christians, and thus we can hardly claim that Christianity (or theistic belief in general) provides a bulwark against toxic philosophies such as Nazism. On the contrary, Nazism fed on the vast pool of Christian anti-Semitism existing at the time – anti-Semitism that stemmed ultimately from the fact that the Jews denied the divinity of Jesus, coupled with the dogma that they were responsible for Christ's death. A good case could be made that it is in Christian anti-Semitism that we find the real roots of the Nazi atrocities.

It's only natural

Let's put aside these specific cases and delve into some of the general issues raised by the above discussion. Several of the alleged ethical implications we've considered so far rest on a hidden assumption: the idea that if something is natural, it must be good. This is an extremely pervasive view, and it forms the basis of a highly influential school of thought in moral philosophy, known as *natural law theory*. According to natural law theorists, if something is natural it is morally acceptable, whereas if it is unnatural it is morally unacceptable. This formula is most commonly employed in the arena of sexual behaviour. According to traditional natural law ethicists, the biological (read: God-given) purpose of sex is procreation, and therefore sex is only

---

[31] Isaac (2005).

morally permissible when used for that purpose. Penises and vaginas are clearly designed for heterosexual intercourse and reproduction, and to use them in ways other than this is an insult to the God that designed them.[32] This idea underlies much of the opposition to homosexuality.[33] A lot of people are closet natural law ethicists, and use natural law-style arguments without even realizing that that's what they're doing. For example, some argue that it is unnatural to remain voluntarily childless, or to commit suicide, or to genetically modify food, and thus that these things are morally wrong. Some argue that eating meat is natural and therefore permissible; others argue that it is unnatural and therefore impermissible.[34] Many people are willing to change society (supposedly not a part of nature), but brand efforts to alter our genetic makeup 'playing God'. All of these arguments are natural law arguments.

The is-ought fallacy does not rule out the possibility that we can assume that what is natural is good or that what is unnatural is bad. Nonetheless, there are good reasons to discard this moral principle. Reason number 1: the terms *natural* and *unnatural* have an evaluative component. To prove this to yourself, imagine that someone told you that 'modern humans keep themselves unnaturally clean'. You would be more likely to assume that this person was opposed to modern cleaning practices than you would be if the person said 'modern humans keep themselves cleaner than did humans in our ancestral environment'. To call something 'natural' is at once to say something (vague) about its

---

[32] See, e.g., Aquinas (1948).

[33] Even if we accept the basic logic of the natural law argument against homosexuality, we have to ask just how wicked a sin it could really be. People sometimes use metal coat hangers as impromptu TV aerials, a purpose for which they were not designed. Likewise, children sometimes climb up slides instead of sliding down them. Are these activities heinous infractions of the moral law? Are they an insult to the people who designed the coat hangers or the slides?

[34] The worst example I've come across of the former argument was in a letter to *Time* magazine: 'If animals aren't intended to be eaten, why are they made of meat?' I'm not sure if the writer was serious.

origin and to commend that thing. To the extent that the word *natural* has positive connotations, arguing that 'X is good because it's natural' is circular; it is, in effect, arguing that X is good because it's good.

Even if we strip the word *natural* of its positive connotations, it would be impossible to establish that *natural* means *good*. I've already hinted at the fact that some things we know to be natural are far from desirable. Aggression is a good example. Most people these days think that aggression is bad, even though it is a natural human propensity. I suppose we could argue that people are wrong to view things this way; maybe we should say that, if the capacity for aggression evolved, aggression must be morally acceptable. But what reason do we have to accept this? If we take the idea seriously, and look carefully at what it implies, we arrive at some quite ludicrous conclusions. Is aggression morally acceptable only when used to harm people (its natural function), and not when focused on an inanimate target as a way of letting off steam? Is the angry husband who hits a punch bag instead of hitting his wife behaving immorally in displacing his aggression in this unnatural way? In the end, there are no final, objective answers to these kinds of questions. It all comes down to a decision. We've got to decide which is more important to us: the notion that the harm caused by aggression is bad, or the notion that what's natural must necessarily be good. I know which option I prefer.

Let's dispense, then, with the idea that *natural* equates to *good*. Next we need to examine the idea that *unnatural* equates to *bad*. It often seems that branding something unnatural is sufficient to establish that it is morally wrong. It's not. For the argument to work, we must establish, first, that the behaviour in question really is unnatural, and, second, that this implies that it is bad. Many natural law arguments fall at the first hurdle: they claim something is unnatural but it turns out not to be. Homosexuality, for instance, is natural, at least in the sense that it is found in other species and that there is almost certainly a genetic contribution to its development. (There are non-genetic factors at work as well, of course.) However, even if a class of behaviour *is* unnatural, we have yet to show that 'unnatural' equates to 'bad' or 'wrong'. Assume for a moment that homosexuality is unnatural. So what? In many other contexts, we are completely at ease with deviations away from what's natural. In his essay *Evolution and Ethics*, Thomas Huxley noted that, when a gardener cultivates a garden from a pre-existing wilderness, we do not consider this immoral or believe that the garden is inferior to the wilderness.[35] Admittedly, many people today would nominate the wilderness state as the more desirable. But you get the point. A lot of things we consider good are unnatural.

In case you didn't like Huxley's example, I'll give you some of my own. It is unnatural to die of 'natural causes'. Most animals, including human beings in our natural environment (i.e., the environment of our pre-agricultural ancestors), die not of so-called 'natural causes', but of injury, disease, or predation. In high-tech societies, people live unnaturally long lives. In parts of the world where people have adopted good hygiene practices, unnatural levels of cleanliness have led to unnatural levels of good health. We keep our teeth unnaturally clean and have an unnaturally low rate of death from dental decay. Pain-relieving drugs are unnatural but most of us are happy to use them when we've got a headache or when we're undergoing surgery. You can probably think of other examples yourself. What all these examples show is that there is nothing inherently wrong (or right) with interfering with the 'natural' order of things. And we must remember that our capacity to interfere is itself part of the natural order. (If that sounds to you like a *justification* for interfering, you haven't digested the central point of this section. Read it again!)

Certainly, people can point to cases where the natural option is better and more desirable than the alternative. A diet rich in natural foods is better than one rich in artificial foods, for instance. However, it's better because it happens to be healthier (at least at the time of writing), *not* because it's natural per se. If it were healthier to eat

[35] Huxley (1894).

artificial foods, then surely we should eat them in preference to natural ones. Or should we poison ourselves with natural foods just because they're natural? Similarly, if genetically modifying food is wrong, this is because it's harmful or potentially harmful, *not* because it's unnatural. And if it's not harmful, then surely it's not wrong. Our decisions should be based on an assessment of real risks and benefits, not on half-baked ideas about naturalness v. unnaturalness.

An evolutionary perspective undermines the worldview in which we can assume that what is natural is good. Natural law theory only makes sense within the framework of a belief that God created all things, and that he is entirely and completely good. To the degree that evolutionary theory undermines the existence of a good God, it pulls the rug out from underneath natural law theory. As soon as we realize that life came about through an amoral natural process, the dogma that what is natural is good goes straight out the window. If anything, we might find ourselves agreeing with Huxley's suggestion that, to be truly moral, it is often necessary to go *against* the flow of nature. 'The ethical progress of society', he wrote, 'depends, not on imitating the cosmic process and still less on running away from it, but in combating it.'[36] Amen to that.

The devil in the details

Imagine you've weighed up all these arguments, but still can't shake the conviction that, if a given behavioural tendency has an evolutionary origin, this somehow justifies it – that what natural selection has 'designed' us to do is what we ought to do. The arguments we've considered so far are, in my humble opinion, strong enough to kill an elephant, but if you're still not persuaded, that's fine. The next step is to stop resisting the idea and see what happens if we actually sit down and make a sincere effort to derive a code of conduct from an analysis of our evolved behavioural dispositions. Is it even possible?

[36] Huxley (1894), p. 83.

An initial problem is deciding which of our evolved preferences or dispositions we should follow. We cannot follow all of them; after all, our various impulses often come into conflict with one another. For example, as we've already seen, our altruistic impulses have their origin in natural selection, but so do our aggressive impulses. We need some principle to adjudicate between these competing drives. The philosopher Robert Richards suggests that this principle should be to favour whatever is best for the group.[37] Altruism is good for the group and aggression is not, and thus we should nurture our evolved altruistic tendencies and seek to control or sublimate our evolved aggressive impulses. This suggestion clearly has some merit, *but the principle is not derived from an analysis of our evolved nature*. We have abandoned the attempt to derive our morals from evolutionary psychology.

Next consider some of the difficult questions that crop up when we try to derive a code of conduct from an understanding of our evolved nature. Does the fact that we evolved to be somewhat altruistic imply that we ought *only* to be somewhat altruistic? Does it imply that it would be wrong to be less altruistic, but also wrong to be *more* altruistic? We're not all genetically disposed to the same level of altruism. So should we all be no more and no less altruistic than the average? Should those below the adaptive optimum take classes in being more altruistic, while those above the adaptive optimum take classes in being more selfish? Or should we each be only as altruistic as we're genetically inclined to be (as if it's even possible to disentangle the genetic from the environmental contribution)? Is it morally wrong to encourage children to be more altruistic than this? Are the Jesus Christs and Ghandis of the world immoral for extolling us to engage in an unnatural level of kindness and generosity?

Another, even more serious problem with deriving a system of ethics from an analysis of our evolved nature is that it would lead to an unworkable and logically inconsistent ethical system. A woman

---

[37] Richards (1999).

who cuckolds her partner may enhance her own fitness (because women who cheat on their partners often cheat with males who are more physically attractive, and physical attractiveness seems to be indicative of superior genes). At the same time, though, she will diminish her partner's fitness (because he ends up investing his biological resources in a child that isn't his own). Let's assume that evolution tells us what's right and wrong. What would it tell us in this situation? It would tell us, first, that it was right for the woman to cuckold her partner (because it enhanced her fitness), and second that it was wrong for her to cuckold her partner (because it decreased *his* fitness). In other words, it would give us logically inconsistent moral counsel. We couldn't implement such a moral system even if, for some bizarre reason, we wanted to.

I'll give you another example. Imagine that a young man and woman have become romantically entwined and that, despite their best intentions, she has fallen pregnant. If he's an attractive fellow, skilled in the art of seduction, his fitness might be best enhanced by abandoning her and finding another woman to impregnate.[38] Her fitness, on the other hand, might be best enhanced by having him stick around to help raise the child, at least for the first few years. What is the morally correct course of action for the man in this situation? Again, evolutionary theory would give us logically inconsistent moral advice. From the point of view of *his* fitness, he should abandon her; from the point of view of hers, he should not. Needless to say, most people would say that the man would be wrong to abandon her purely so that he could impregnate another woman. But this judgment doesn't come from evolutionary theory. Again, we are using our pre-existing moral standards to appraise the situation.

---

[38] Actually, the best thing a man could do to enhance his fitness in the modern environment would be to spend as many of his waking hours as possible donating sperm to sperm banks around the world. But it would be a strange ethical code that mandated that particular course of action! So there's another good reason to reject the idea that evolutionary theory implies we should act to enhance our fitness.

And if we can do that, why do we think we need to derive our morals from an analysis of evolved dispositions? We already *have* our morals!

What Darwin contributes to ethics

Many of the ethical conclusions drawn from evolutionary theory in the past are without foundation. Nonetheless, the theory does have a contribution to make to ethical discourse. As we saw earlier, although we cannot derive our values directly from evolution, the theory highlights certain facts about the world which, when combined with our pre-existing values, may furnish us with novel moral recommendations. Before wrapping up this chapter, we'll consider a few examples. One is based on Robert Trivers' theory of reciprocal altruism, which was introduced in Chapter 11. According to Trivers, a potentially evolvable behavioural strategy is: 'first do unto others as you wish them to do unto you, but then do unto them as they have just done unto you'.[39] If we were to assume that this is the right way to act simply because it is what has evolved, we would be committing the is-ought fallacy. But there's another way to proceed. Reciprocal altruism theory highlights the fact that, if you give to others even if they do not give to you, you're liable to be exploited. Assuming you don't want to be exploited, you might decide to reject moral precepts such as 'turn the other cheek' and 'resist not evil', and instead deny help to those who don't reciprocate. On the other hand, if for some reason you *do* want to be exploited, you might decide to help regardless of whether people reciprocate or not. Better yet, you might extend help selectively to those who do *not* reciprocate. Evolutionary theory cannot tell you whether it's better to be exploited or not. However, once you've made that choice (and obviously it's not a hard choice to make), it can inform your decisions about how to act.

Evolutionary psychology also has implications for social policy.[40] Guided by evolutionary principles, psychologists Martin Daly and

---

[39] Trivers (1985), p. 392.
[40] For an overview of the topic, see Crawford and Salmon (2004).

Margo Wilson discovered that one of the best predictors of homicide rates in a given area is the degree of economic inequality in that area.[41] That is, the greater the gap between the richest and the poorest, the greater the homicide rate. Absolute levels of poverty don't matter, just the size of the gap. Daly and Wilson suggest that an obvious policy recommendation stemming from this finding is that we ought to reduce economic inequality. (It is ironic, in light of such a suggestion, that sociobiology has so often been accused of bolstering right-wing political views.) It's important to note, though, that the policy recommendation does not follow *directly* from Daly and Wilson's discovery. Reducing economic inequality is recommended only if we wish to lower the homicide rate, and if we consider this more important than maintaining the sort of sink-or-swim society that fosters economic inequality. If we don't wish to lower the homicide rate, or – more realistically – if we think that a somewhat elevated homicide rate is a price worth paying for the benefits of a sink-or-swim society, Daly and Wilson's finding has no such implication. But because most people would like to see the homicide rate lowered at any cost, this may not be obvious. We jump straight from an *is-statement* ('inequality leads to a higher homicide rate') to an *ought-statement* ('we ought to reduce inequality'), without articulating the implicit intervening premise that justifies this leap: that it is desirable to try to remove any conditions contributing to an elevated homicide rate. Research in evolutionary psychology can legitimately inform our decisions about how to behave and how to organize society, as long as we can justify the leap from *is* to *ought*.

This raises another point. Often when we come across an argument that seemingly derives an ethical conclusion directly from a fact about evolution, there turns out to be an intervening premise lurking somewhere in the background – an implicit premise that makes the argument deductively valid. Several examples can be drawn from the

---

[41] Daly *et al.* (2001).

writings of E. O. Wilson. First, according to Wilson, certain facts about evolution imply that one of our central values should be the protection of human life.

> If the whole process of our life is directed toward preserving our species and personal genes, preparing for future generations is an expression of the highest morality of which human beings are capable.[42]

I think we can agree that preparing for future generations is a highly desirable value to cultivate. Surely, though, this is *not* because the 'process of life' is directed towards the preservation of our groups or genes; it is because suitable preparation will secure a better and happier life for the future inhabitants of this planet. Although Wilson doesn't make this clear, I assume this is the real reason he thinks we should prepare for future generations: it will increase happiness and reduce suffering. Thus his argument is not a deduction from the facts of evolution; it is a utilitarian argument.[43] And when you scratch beneath the surface, it turns out that many of the arguments of the evolutionary ethicists rest on utilitarian considerations, rather than on the simple fact that a given trait evolved or that a chosen goal is consistent with the overall direction of evolution.

Another example comes from another of Wilson's claims: that evolutionary biology shows that we are dependent on the rest of the biosphere, and that this implies that we have a duty to protect and preserve the human gene pool and the diversity of life on the planet.[44] Again, I would happily endorse Wilson's *conclusion*, but again it is not a deduction from the facts of evolution. It involves taking a biological fact – our dependence on other life – and conjoining it with a pre-existing

---

[42] Wilson (1984), p. 121.

[43] The basic position of the utilitarian can be summed up like this: of any two things that someone could do at a given time, the morally superior one is the one that we can reasonably expect to have the best overall consequences for all affected. Usually this means the act that produces the most happiness or pleasure, or that alleviates the most unnecessary suffering.

[44] Wilson (1984).

ultimate value: that human life is valuable and should be preserved. Without that value, the argument is deductively invalid. Certainly, evolutionary biology highlights the intricate interdependence of all life on this planet, and in doing so makes us aware that, if we wish to maintain human life, we need to maintain biodiversity. However, it does not imply that we *must* preserve biodiversity. That's up to us. As soon as we decide what we value in life, evolutionary theory can help guide our actions and decisions. But until we make that decision, the theory is ethically inert.

Conclusion

People often assume that anyone who studies evolution thinks that everything should be about the survival of the fittest. But this makes no more sense than assuming that anyone who studies glaciers thinks that everything should be done really, really slowly. We have no reason to mimic the way of nature, and no reason to assume that something is good simply because it is a product of evolution. To explain the origin of something bad is not to show that it is actually good, even if the explanation invokes natural selection. Natural selection accounts for some things that we consider good and some that we consider bad. It is not a font of moral wisdom or a source of ethical advice.

Most evolutionary psychologists are aware of this, and have bent over backwards to avoid the accusation that they are promoting any of the undesirable behaviours they claim are products of evolution (e.g., a tendency to stray sexually). Indeed, it could be argued that they've bent over too far and have become overly moralistic, sacrificing scientific neutrality for the active promotion of a moral code opposite to the one the critics have accused them of promulgating. Still, this might be unavoidable. Critics have told people that evolutionary psychology is false but that, if it were true, it would have all sorts of terrible implications. What happens if people decide that the critics are wrong about the first part (that evolutionary psychology is false), but don't work out that they're wrong about the second as well? The danger is that they'll accept the implications outlined by the critics. As such, it may be incumbent on evolutionary psychologists to stress and stress again that these implications simply do not follow from the truth of an evolutionary explanation for human behaviour. Evolutionary theory does not justify the status quo, or imply that inequalities between the sexes or between different social groups must be preserved. Evolutionary theory does not imply that people can never be held accountable for their actions or that their behaviour can never be changed or improved. Evolutionary theory does not imply that unrestricted capitalism is the only acceptable political system or that social welfare should be abolished. And evolutionary theory does not imply that aggression or other undesirable traits are acceptable simply because they are natural.

This is where the discussion of evolution and ethics often ends. However, there are a number of important ways in which Darwin's theory can inform moral discourse. As we've seen, it brings to our attention facts about the world that, when combined with our pre-existing ethical values, can provide us with new ethical insights. On top of that, it undermines support for some traditional moral theories, in particular natural law theory. Thus evolutionary theory and evolutionary psychology do threaten to overturn some of our most cherished values after all. In the last two chapters of the book, we'll take this idea and see how far it can be pressed.

THIRTEEN

Uprooting the doctrine of human dignity

Animals – whom we have made our slaves we do not like to consider
our equals. – Do not slave holders wish to make the black man other
kind?

Charles Darwin,
cited in Barrett *et al.* (1987), p. 228

Few people seem to perceive fully as yet that the most far-reaching
consequence of the establishment of the common origin of all species
is ethical; that it logically involves a readjustment of altruistic morals,
by enlarging, as a necessity of rightness, the application of what has
been called 'The Golden Rule' from the area of mere mankind to that
of the whole animal kingdom. Possibly Darwin himself did not quite
perceive it.

While man was deemed to be a creation apart from all other
creations, a secondary or tertiary morality was considered good
enough to practise towards the 'inferior' races; but no person who
reasons nowadays can escape the trying conclusion that this is not
maintainable.

Thomas Hardy, 1910,
cited in Millgate (2001), p. 311

All the arguments to prove man's superiority cannot shatter this hard
fact: In suffering the animals are our equals.

Peter Singer (2002b)

Putting morality on a new foundation

A lot of the resistance to Darwin's theory of evolution by natural selection is based on the idea that it undermines traditional morality and that this is a bad thing. Among those who accept the theory, there are several responses to this. Some agree that it would indeed be a bad thing if evolution undermined morality, but argue that, fortunately, it does not. Others, though, have drawn a different conclusion. They agree with the anti-evolutionists that Darwin's theory undermines traditional morality, but, unlike the anti-evolutionists, they think that this is potentially a very good thing. Evolutionary theory may prompt us to devise a new and superior morality, a morality that supersedes pre-Darwinian ethical codes. In the last chapter, we considered a number of failed attempts to derive ethical implications from evolutionary theory. In this chapter, we'll consider a more successful attempt – one that is consistent with a sophisticated understanding of the theory and that has important implications for various topics in applied ethics. These include suicide, euthanasia, and the proper treatment of non-human animals.

But before plunging into these heated issues, a brief note of caution. In the eighteenth century, the philosopher Edmund Burke argued that we should not tamper too freely with long-standing social institutions, including the institution of morality. The fact that these institutions have survived for long ages suggests that they must be doing something right. This is, in effect, a Darwinian argument, although of course it predated Darwin. A similar argument, but one based explicitly on Darwinian reasoning, was put forward by the American sociologist William Sumner in the late nineteenth century.[1] Echoing Burke, Sumner noted that we should be wary of meddling with our traditions and customs and practices, as these things are products of a selection-like process taking place over countless generations, and they therefore

---

[1] Sumner (1883).

embody the 'wisdom' of centuries of trial-and-error experimentation. In a sense, our traditions – morality among them – have more wisdom than we do. Just as we could not design a better body than the one natural selection has bequeathed us, nor could we design a better society, or a better system of morals, than the one we've been bequeathed by tradition. Just as our bodies work well even though we lack a complete understanding of how, so too our moral systems work well even though we do not always know how or why. Some of our moral practices and other social institutions may seem silly or pointless to us, but that's because we're not smart enough to discern what functions they serve. Societies are immensely complex, and with our limited intellects we may fail to appreciate how particular institutions contribute to the smooth running of society and the maintenance of peaceable relations among individuals and groups. We may dismantle them because they appear unjustified and arbitrary. But we do so at our own peril. We should think long and hard before casting them aside.

There is some merit in this argument. However, there are three things to bear in mind. First, it's possible that the wisdom latent in widespread social arrangements and moral codes is now obsolete. These things first evolved in a very different world from the one we now inhabit, and thus they may no longer work in the way they were designed to. Second, even if existing social institutions *do* work, we must ask: whom do they work for? Our moral conventions do not necessarily aid and abet everyone. Although it's true that any institution that persists for a long time must have attributes that allow it to persist, this doesn't mean that the institution is good for all members of society. Technically, it doesn't even mean that it is good for *any* members of society! It may simply have attributes that are good for the institution itself – attributes that facilitate the institution's continuation even at the expense of its members.[2] There is therefore no reason to think that institutions such as morality can be justified on

---

[2] Cf. meme theory; see Blackmore (1999); Dawkins (1989); Dennett (1995, 2006).

the basis of their stubborn persistence. A third and final point: although we could not have designed the human body from scratch, we have learned to make beneficial changes to it. For instance, we use laser surgery to reshape our corneas and thereby correct poor eyesight, and we remove one another's appendixes when they become inflamed. Presumably we can make comparable tweaks and adjustments to our moral codes and practices. With this in mind, and with suitable circumspection, let's now consider how evolutionary theory might prompt us to rewrite traditional morality.

The doctrine of human dignity

The pivotal concept in this chapter is an important trend in traditional Western moral thinking, which the philosopher James Rachels dubbed the *doctrine of human dignity*.[3] Peter Singer uses the phrase *sanctity of human life* to refer to essentially the same thing.[4] Although this doctrine is often not explicitly expressed, it is the heart and soul of the Western moral system, and provides the moorings for traditional morality. The doctrine has two parts; the first pertains to humans, the second to non-humans. The part pertaining to humans is the idea that human life has supreme worth; according to some, it has literally infinite value. For this reason, any activity that involves taking a human life (or at any rate, taking an *innocent* human life) is utterly forbidden. This includes suicide, euthanasia, and abortion. The flipside of the doctrine of human dignity is the idea that the lives of non-human animals have vastly less value than the lives of human beings. In fact, according to some commentators, such as the German Enlightenment philosopher Immanuel Kant, they have no inherent value whatsoever. This means that, although it would be wrong for someone to torture your cat, this is not because the cat wouldn't like

---

[3] Rachels (1990).
[4] Singer (2002a).

it; it is because *you* wouldn't like it. Animals exist solely for our benefit and may be sacrificed for our purposes. As Kant put it, humans are an end in themselves, whereas animals are merely a means to whatever ends we choose to pursue.

Evolutionary theory does not directly contradict the doctrine of human dignity. It does something else, though: it undermines the foundations upon which it rests, and the worldview within which it makes any sense. I'll explain how. The doctrine of human dignity has its historical roots in the anthropocentric, pre-Darwinian view of the universe. According to Rachels, it is supported by two main ideological pillars: the image-of-God thesis and the rationality thesis. The image-of-God thesis is the idea that we and we alone were created in the image of God and have an immortal soul, and that this is a fundamental difference between all humans and all other animals that justifies a fundamental difference in how we *treat* all humans v. all other animals. The rationality thesis is the idea that we and we alone possess rationality – the spark of reason – and that *this* is the difference that justifies privileging members of our species above all others.

Evolutionary theory undermines both these theses. Let's start with the idea that we are made in the image of God. Given that the raw materials with which natural selection works are a product of random mutations and the random recombination of genes during sexual reproduction, and given also that evolutionary history is shaped by completely unpredictable external events (such as asteroids smashing into the planet, which is what wiped out the dinosaurs) – given these facts and the randomness inherent in evolution, it becomes very difficult to believe that human beings were created in the image of God or in accordance with *any* pre-existing design. Admittedly, it is possible in principle. One could argue, for instance, that God somehow set up the process of evolution in such a way that it was guaranteed to manufacture a species in his psychological likeness, or that he personally guided the process with this goal in mind.

However, even if we ignore the desperate ad hoc flavour of these attempted fixes, we've already seen that it is very difficult to reconcile evolutionary theory with the existence of God. At a push, the theory is consistent with the existence of an evil God (Chapter 6) or a non-anthropomorphic 'God' (Chapter 7). But neither of these conceptions is capable of grounding the image-of-God argument for human dignity. If God were evil, the fact that we were made in his image would no longer imply that we deserved special moral treatment; if anything, it would imply the opposite. And if God were some weird abstract principle, we could not have been made in God's image. What could it mean, for example, to say that we were made in the image of the Ground of All Being?

So much for the image-of-God thesis. Next, let's consider the rationality thesis: the idea that we are distinguished in some morally significant way from other animals by our possession of rationality. As mentioned in Chapter 8, we now know that the ability to reason (i.e., to work things out about the world) is not found only in humans. Certainly, we have a bigger helping of it than any other animal; however, reason or rationality is distributed in differing degrees throughout the animal kingdom, and thus the difference between us and other animals is one of degree rather than of kind. If people possess full rationality, then other apes and other animals possess at least partial rationality.[5] So even if rationality did confer moral worth, we would still have no justification for allocating moral worth to all human beings but not to any other animal. But even if we really were the only animals possessing rationality, it wouldn't matter because, as we saw in Chapter 9, from an evolutionary perspective, the capacity for reason is simply one adaptation among many millions that have evolved in the animal kingdom. We like to think that reason is the supreme adaptation; that rational animals deserve preferential

---

[5] Note that it is easy to imagine a more rational evolved being than *Homo sapiens*, and in relation to that being we ourselves would only be partially rational.

treatment; and that non-humans, because they don't have reason, have no intrinsic moral value. However, after Darwin, this is no different and no more convincing than, say, an elephant thinking that *trunks* are the supreme adaptation; that animals with trunks deserve preferential treatment; and that non-elephants, because they don't have trunks, have no intrinsic moral value. If we came across this view, we'd instantly dismiss it as an elephant-biased view of the world. However, the idea that reason is the supreme adaptation is an equally *human*-biased or anthropocentric view of the world.

By undermining both the rationality thesis and the image-of-God thesis, the Darwinian worldview undermines the doctrine of human dignity. It leaves it without intellectual foundations. This has important implications for many key issues in ethics. The idea that human life – and human life alone – is infinitely valuable has impregnated the ethical systems of the world, especially those of the West. Although the doctrine of human dignity has its origins in the religious conception of humankind, it has woven its tendrils into our secular codes of ethics. It is implicit in the ethical beliefs of many who doubt or even reject the various religious accounts of human origins, and who believe that right and wrong exist independently of religion. Thus, even though we in the West live in a semi-post-Christian world, in which the image-of-God thesis and the rationality thesis are widely dismissed, the ethical attitudes they inspired linger on. But what happens to these attitudes when we really get to grips with the fact that the foundations of our traditional morality have eroded? That is the question we'll examine in the remainder of this chapter.

Suicide and voluntary euthanasia

We'll start with the vexed and disconcerting issue of suicide, and the closely related topic of voluntary euthanasia (also known as assisted suicide). Philosophers have debated the ins and outs of self-killing for thousands of years. The questions they ask are provocative. Are we obliged to stay alive if we really do not want to? Should people have the right to kill themselves? Should people have the right to *stop* others from killing themselves, if that's what they really want to do? As a general rule, philosophers and religious thinkers have been opposed to suicide. Christianity has taken a particularly hard line. Saint Augustine, for example, taught that suicide is unacceptable in all circumstances, always. In his view, it is not merely a garden-variety sin but an *unrepentable* one: a sin that God will never forgive. Similar attitudes are found in other religions. Suicide is generally frowned upon in Islam, for example, although Muslim opinion differs on the permissibility and desirability of suicide bombings and related activities. Even in cultures and times in which suicide has been judged less harshly, or even considered mandatory in certain circumstances, it is generally held that suicide is shameful and morally undesirable.[6]

Such attitudes have been supported by a number of arguments. Some of these are no longer viable in a post-Darwinian world. This includes such arguments as that suicide is unnatural; that it's up to God to choose the moment and the manner of our deaths; and that God has forbidden us from taking our own lives. As we know, evolutionary theory strongly challenges the existence of God. This immediately weakens any appeal to God's authority. It also undermines any appeal to the supposed unnaturalness of suicide; as we saw in the last chapter, natural law arguments only make sense within the context of

---

[6] Although most thinkers have opposed suicide, there are a few interesting exceptions. Friedrich Nietzsche (2003), always the iconoclast, wrote: 'The thought of suicide is a powerful solace: by means of it one gets through many a bad night' (p. 103). And then there was Hegesias of Cyrene, a little-known ancient Greek philosopher, who argued that, in most people's lives, misery outweighs pleasure and thus that we should all commit suicide. This earned him the cuddly little nickname *Peisithanatos*: the death-persuader. It is not known whether he ended up taking his own advice. But it's hardly surprising that his philosophy is virtually unknown today. A philosophy whose most enthusiastic advocates kill themselves, and are therefore not around to spread the word, is almost certainly fated for extinction.

a theistic worldview. There's another argument against suicide, though, and for present purposes it's the most important. It is based on the doctrine of human dignity. A natural corollary of the idea that human life is infinitely valuable is that taking a human life – including one's own – is infinitely wicked. Thus suicide is wrong for the same reason that murder is wrong: because human life is sacred.

A similar argument has been made in regard to voluntary euthanasia. Kant noted that when an animal is suffering we put it out of its misery, and that's OK, but it's not OK when it comes to human beings because of the infinite worth of human life. Similarly, the Rabbi Moshe Tendler opposed voluntary euthanasia on the grounds that 'Human life is of infinite value'. In his view, we should not cut short a person's life by even a few days because 'a piece of infinity is also infinity, and a person who has but a few moments to live is no less of value than a person who has 60 years to live'.[7] In short, the injunction against assisted suicide – like that against unassisted suicide – is commonly underwritten by the doctrine of human dignity.

But the whole edifice starts to crumble once we bring Darwin into the picture. With the corrective lens of evolutionary theory, the view that human life is infinitely valuable suddenly seems like a vast and unjustified over-valuation of human life. And if human life is *not* supremely valuable after all, then there's no reason to think that suicide or voluntary euthanasia is necessarily wrong in all circumstances. In fact, it starts to seem decidedly odd that we have elevated human life – i.e., pure biological continuation – so far above the *quality* of the life in question for the person living it. Why should life be considered valuable in and of itself, independently of the happiness of the individual living that life? Needless to say, we must be very cautious with this argument, especially when it comes to suicide. Most people who kill themselves have not thought their decision

---

[7] Cited in Rachels (1990), p. 204.

through properly, and if they'd managed to ride out the suicidal crisis, they would have had perfectly good and happy lives. Many suicidal individuals are severely depressed, and severe depression involves an unrealistically negative apprehension of the future and the hopelessness of one's situation. Rational suicides – suicides based on an accurate picture of one's situation and future prospects – are comparatively rare. Furthermore, in assessing the rightness or wrongness of suicide, we need to take into account its effects on those left behind, as suicide usually causes immeasurable grief and suffering to the victim's family and other loved ones. Nonetheless, after Darwin, it is difficult to maintain an absolute prohibition on suicide. There may be circumstances – rare and unhappy circumstances – in which suicide is a reasonable and ethically permissible course of action. In any case, this possibility cannot be ruled out on the grounds that human life is infinitely valuable.

The argument is even stronger when it comes to voluntary euthanasia. If life is not infinitely valuable, there is no reason to assume that the duty to preserve human life should always take precedence over other considerations, such as human happiness and the avoidance of suffering. Thus voluntary euthanasia is no longer ruled out as an absolute evil. As with suicide, there may be circumstances in which we decide as a society that it is morally permissible. We may decide, for instance, that euthanasia is an acceptable course of action when someone with a painful terminal disease has a persisting, rational, and non-coerced desire to die – even though it involves taking an innocent human life.

Critics of euthanasia argue that it is immoral to take a person's life, even when that person is suffering and wishes to die with dignity. After Darwin, however, we might be more inclined to argue that it is immoral to force people to keep on living when they would rather not. Here's something to think about. In many ways, we treat other animals abysmally. But if a horse or a dog or a cat is suffering terribly from a terminal injury or disease, or if it has limited prospects for quality of life in the future, most people agree that the humane thing to do is to put it out of its misery. Not to do so would be considered inhumane. However, because of the inflated value traditionally assigned to human life, we are *less humane* in our treatment of *human beings* who are suffering or have a painful terminal illness. This is an ironic exception to the general rule that the doctrine of human dignity secures better treatment for humans than for non-humans. In this one instance, we treat non-human animals more humanely than we do human beings, because of superstitious beliefs about the value of human life. People are made to suffer needlessly because of superstition. If it is acceptable to put non-humans out of their misery, why is it not acceptable to do the same for people who request it, or in some cases beg for it? One might even argue that euthanasia is *less* morally problematic in the human case, because people can give their explicit and reasoned consent, whereas other animals cannot.

An evolutionary perspective doesn't completely solve the problem of suicide or euthanasia. Many questions remain. In what circumstances should we forcibly prevent people from killing themselves? Should physician-assisted suicide be available to people who are simply tired of life? Evolutionary theory cannot answer questions like these. What it can do, though, is deactivate some of the traditional arguments against suicide and euthanasia, removing them from the table and thereby opening up the possibility that, at least in some situations, they might be acceptable courses of action. If nothing else, evolutionary theory wakes us up to the fact that we can no longer take a ready-made answer to these difficult questions from a holy book or religious authority. We're going to have to think it through.

Of course, an evolutionary perspective does not imply that we should take suicide or euthanasia lightly. On the contrary, a strong argument can be made for the opposite view. As we saw in Chapter 6, the evolutionary process that gave us life involved the suffering of untold millions of people and other animals. Does this not oblige us to cherish our existence if we possibly can, to make the most of the life that our forebears unwittingly bequeathed us with their torments and agonies?

The subjection of animals

A second set of implications relates to the proper treatment of non-human animals. As we've seen, the doctrine of human dignity assigns non-humans to a lowly position on the scale of life. According to Saint Thomas Aquinas, animals exist for the sake of humans, not for their own sake, and thus 'It is not wrong for man to make use of [animals], either by killing or in any other way whatsoever.'[8] This is a sentiment many people have lived by. Historically, and even today, we have treated other animals appallingly. This is one of my favourite quotations; it comes from the theologian W. R. Inge:

> We have enslaved the rest of the animal creation, and have treated our distant cousins in fur and feathers so badly that beyond doubt, if they were able to formulate a religion, they would depict the Devil in human form.[9]

In the nineteenth century, Karl Marx argued that, throughout history, the ruling classes have stolen the labour and energy of the labouring classes. It could also be argued that, in precisely the same way, human beings have stolen the labour and energy of the animals they have forced to work for them. Civilization is built on the blood, sweat, and tears of our animal cousins. A number of commentators have gone so far as to liken our treatment of the animals to the Nazi Holocaust.[10] As Isaac Bashevis Singer, winner of the Nobel Prize for Literature, noted in one of his short stories:

[8] Aquinas (1928), III, II, p. 112.
[9] Inge (1922), pp. 166–7.
[10] Patterson (2002).

They have convinced themselves that man, the worst transgressor of all the species, is the crown of creation. All other creatures were created merely to provide him with food, pelts, to be tormented, exterminated. In relation to them, all people are Nazis; for the animals it is an eternal Treblinka [the Nazi extermination camp].[11]

A lot of people find the comparison between the Holocaust and our treatment of animals insensitive and deeply offensive. However, the comparison is only offensive if we've decided already that the lives of human beings in the Holocaust are vastly more important than the lives of other animals in comparable situations. And that's what the debate is all about! The point of the comparison is not to trivialize the Holocaust; it is to suggest that the plight of animals is being trivialized. If people want to claim that the comparison is erroneous or inappropriate, they must explain *why*. Simply to complain that it is offensive is to *assume* that it is erroneous without first justifying the assumption.

Let's look at what an evolutionary perspective contributes to the debate. First of all, some suggest that Darwin's theory justifies our domination of the animals – it is a manifestation of the survival of the fittest, the struggle for existence, the law of the jungle. According to this argument, the fact that we *can* dominate the animals in itself justifies our domination of them. Now it is true that we don't have to be nice to animals; they're at our mercy. But we can hardly argue that our actions are morally acceptable just because we can get away with them. That said, if you think that our ability to dominate other animals is sufficient justification for the way we treat them, and if you really don't care about their suffering, there won't be much I can say to change your mind. However, you should at least be aware that your argument has the same underlying structure as the argument that, because bullies are able to dominate their victims, bullying is justified. Not only that, the same argument could be used to

---

[11] I. B. Singer (1968), p. 270.

demonstrate that, if the Nazis had won the Second World War, Nazi rule would be justified. Are you sure you want to accept this argument?

You should also bear in mind that few if any evolutionists think that evolutionary theory licenses the mistreatment of non-human animals. Indeed, in one of his major works, Darwin himself wrote that 'humanity to the lower animals [is] one of the noblest virtues with which man is endowed'.[12] And it was a virtue that he himself exhibited in his day-to-day life. Darwin's son, Francis, recalled that his father:

> returned one day from his walk pale and faint from having seen a horse ill-used, and from the agitation of violently remonstrating with the man. On another occasion he saw a horse-breaker teaching his son to ride, the little boy was frightened and the man was rough; my father stopped, and jumping out of the carriage reproved the man in no measured terms.[13]

This shows that it is possible to be an evolutionist and to care about the plight of other animals. More than that, an evolutionary perspective may actively promote a concern for animal welfare, by forcing us to rethink the moral status of our 'cousins in fur and feather'. First and foremost, an evolutionary perspective challenges the doctrine of human dignity, with its brute distinction between human beings and all other life. Darwin's theory stresses our kinship with the animals. It reveals that humans are not a special creation or a separate kind of thing. Chimpanzees, dolphins, frogs – these are literally our distant relatives. In practice, it is easy enough to draw a human/animal distinction, and to import this into our moral reasoning. But evolutionary theory shows that this distinction does not have the significance it was once assumed to have. On top of that, the theory undercuts other arguments justifying the exploitation of animals.

---

[12] Darwin (1871), p. 101.
[13] F. Darwin (1887c), p. 200.

This includes the argument that God put the animals here for our sake. Before scientists pieced together a picture of the history of life on this planet, this might have seemed like a reasonable claim. It's not reasonable anymore. We know now that the vast majority of animals finished their sojourn on this planet long before we evolved. We also know that we and the other animals came about through the same natural process, and that our so-called 'creator' (i.e., natural selection) had no special affection for us. In light of these facts, the suggestion that animals are here for us seems self-centred, quaint, and – to be clear about this – patently false. As the Pulitzer Prize-winning African American author Alice Walker wrote, non-human animals 'were not made for humans any more than black people were made for whites or women for men'.[14]

Bearing these points in mind, the practice of allocating moral status to human beings, but denying it to any other animal, starts to look arbitrary and unjustified. Still, there are various ways that people might try to defend it, and we should consider the most important of these. One common approach for the human supremacist is to argue that non-human animals stand outside the moral community and are therefore exempt from moral consideration. This is based on the idea that morality is an unspoken contract of rights and obligations that individuals enter into for their mutual benefit.[15] As members of the moral community, we are granted rights (e.g., the right to life), but we also assume corresponding obligations (e.g., the obligation not to kill). The only individuals who have rights are those that can grant rights to others. Animals are incapable of entering into the moral contract that brings rights into existence, it is argued, and therefore they do not have rights.[16] There is, in my opinion, a

---

[14] In the preface to Spiegel (1996), p. 14.

[15] This is known as the Social Contract approach to ethics and political philosophy. Among the most important proponents of this approach are Hobbes (1651) and Rawls (1971).

[16] Scruton (2000).

major flaw in this argument: whether rights entail obligations is a matter to be decided, not an objective fact about the world. Rights are not really real; they simply represent decisions we've made about how we should treat one another. We can give rights to whomever we want. In the end, it's a matter of taste, and logic cannot compel us to adopt any particular position. However, if I had to choose to live in a world where the ascendant moral principle was 'reduce unnecessary suffering', or 'don't grant rights to anyone who can't return the favour', I know which I'd choose.

Another common argument (or assertion) is that humans only have moral obligations to other humans – in other words, that we should 'look after our own'. The philosopher Robert Nozick put the point like this:

> The members of any species may legitimately give their fellows more weight than they would give members of other species (or at least more weight than a neutral view would grant them). Lions, too, if they were moral agents, could not then be criticized for putting other lions first.[17]

But Rachels points out that exactly the same argument could be used to justify racial discrimination, as we see in this modified version of the argument:

> The members of any race may legitimately give their fellows more weight than they would give members of other races (or at least more weight than a neutral view would grant them). Whites ... could not then be criticized for putting other whites first.[18]

It's the same underlying argument, so if you accept the first version, then to be consistent you must also accept the second – unless, that is, you can find a *principled* reason for making an exception. You cannot simply assert that the argument applies to species but not to races. Why does it? Granted, the boundaries between different racial groups

---

[17] Nozick (1983), p. 29.
[18] Rachels (1990), p. 184.

are fuzzier and more porous than the boundary between humans and other species. But Darwin showed us that species differences and racial differences are not different in kind, only in degree.[19] To make species the moral dividing line is arbitrary. Why should 'our own' be limited to our species rather than to our racial group? Why, for that matter, should it be limited to our species rather than to, say, our taxonomic class (i.e., mammals)?

So, the allocation of moral status to humans and humans alone is unjustified. This is an important conclusion, but it leaves us with a difficult question: after the elimination of the doctrine of human dignity, how *can* we allocate moral status? Where *do* we draw the moral line? We can't just treat all animals equally; no one would accept that the life of an ant is as important as the life of a human or a chimpanzee, or that swatting a fly should be considered an act of murder. But that does not imply that the only place to draw the line is between humans and all other animals. Fortunately, we don't have to choose between these two options. A useful principle here is the idea that individuals should be accorded moral status in proportion to their degree of sentience or, more specifically, their capacity to suffer. Think about it. Why do we consider it morally repugnant to torture human beings? It's not because they're rational, or because they're members of a moral community, or because they're members of the same species as us. It's because it causes them pain and trauma. If it didn't, we wouldn't worry about it. Now if we decide – and this is our decision; it's not imposed on us from above – if we decide that reducing the amount of suffering in the world is a good ethical principle to live by, then it becomes entirely unjustified and arbitrary to extend this principle to human beings but not also to extend it to other animals capable of suffering. Why should the suffering of non-humans be less important than that of humans? Surely a universe with less suffering is better than one with more, regardless of whether the

[19] Does this sound racist? Only if you have a prejudice against other species!

locus of suffering is a human being or not, a rational being or not, a member of the moral community or not. Suffering is suffering, and these other variables are morally irrelevant.[20]

With the avoidance of suffering as our guiding principle, we have a solid and sensible rationale for many of our ethical intuitions. Humans presumably have a far greater capacity to suffer than flies; therefore, it is far worse to harm a human than to harm a fly. Humans probably also have a greater capacity to suffer than chimpanzees – we are locked into tighter emotional bonds, we grieve for longer – and therefore it is somewhat worse to harm a human than a chimp. (Note that if we accept this argument, we must also accept that it would be worse to harm a member of a hypothetical species with a greater capacity to suffer than us than it would be to harm a human.) Admittedly, this kind of sliding-scale morality is much harder to implement than a morality based on a simplistic, black-and-white distinction between humans and all other animals. Where, for example, do we draw the line between animals it is acceptable to kill and those that it's not? Should the killing of a chimpanzee or a dolphin be considered murder? What about the killing of an elephant? Should we punish the mistreatment of goldfish? Using a sliding scale makes us acutely, uncomfortably aware that we might be getting it wrong. However, if we opt for a morality based on a brute human/non-human distinction, we *know* we're getting it wrong – some animals will definitely be treated worse than they should be. Being uncertain about whether we're wrong is at least a step in the right direction.

A lot of people would happily go along with me this far. However, this approach to moral reasoning does have some contentious and counterintuitive implications. Should the lives of a few mice or monkeys be sacrificed in order to save millions of humans? It's a no-brainer, right? Obviously they should. But why shouldn't the lives of a few humans be sacrificed in order to save millions of mice

[20] For an alternative justification for treating animals nicely, see Regan (1983).

or monkeys? If the lives of mice or monkeys have any moral value whatsoever, then presumably the interests of enough of them (a thousand? a million?) would outweigh the interests of a single human being. If not, why not? Another contentious implication is that the life of a human being is sometimes worth less than the life of a non-human animal. One of the great philosophers of our time, Peter Singer, has argued, for example, that the life of a severely brain-injured infant is worth less than the life of a healthy adult chimpanzee, and that it would therefore be worse to kill the chimp than the infant. Similarly, it would be more humane to experiment on an anencephalic human infant (an infant born with little or no cerebral cortex) than it would be to experiment on a healthy, intelligent chimpanzee – or even a healthy rat. This is because the anencephalic infant presumably doesn't experience pain (or anything else), whereas the chimp and the rat do.[21] Such views are utterly incompatible with the doctrine of human dignity, and if these views seem wrong to you, presumably this is because that pre-Darwinian moral outlook is still operative in your thinking. But can you justify it?

Now some commentators would deny that animals other than human beings are conscious entities, capable of suffering or pain, and if they're right, our principle of avoiding unnecessary suffering would lead us straight back to the conclusion that humans alone have moral standing. But an evolutionary perspective drastically lowers our confidence in the Cartesian view that non-human animals are unconscious automata. After all, *we* are conscious beings (conscious automata perhaps), and we came about through the same process as every other animal. This being the case, it seems unreasonable to deny that any species other than our own has consciousness. If we're still not quite convinced of this, we should remind ourselves that we have a vested interest in maintaining our present, exploitative relationship with other animals, and that this could well bias our thinking

[21] Singer (1993).

on the matter. Just as the slave owner might prefer to view his slaves as inferior and unable to live freely, so too we might prefer to think that the creatures we eat, farm, and force to work for us (our non-human slaves) lack sentience and moral value. In the absence of a very strong argument to the contrary, our default assumption should be that at least some other animals can suffer and experience pain just as we can, and thus that they should be accorded due moral respect.

Considerations such as these support the view of many modern moral philosophers that we have drastically undervalued the lives of other animals. When we grant non-human animals the moral standing they deserve, our relationship with them is transformed. For one thing, we recognize that prejudice against other species (*speciesism*) is just as morally abhorrent as any other form of prejudice, including racism and sexism. As Singer wrote:

> Racists violate the principle of equality by giving greater weight to the interests of members of their own race when there is a clash between their interests and the interests of those of another race. Sexists violate the principle of equality by favoring the interests of their own sex. Similarly, speciesists allow the interests of their own species to override the greater interests of members of other species. The pattern is identical in each case.[22]

Speciesism is so widespread and so rarely challenged that it's easy to overlook the fact that it even is a prejudice. To illustrate: according to some religious opponents of evolutionary theory, if we tell people they're animals, they will behave like animals. But imagine if people worried that if we told white people they're the same type of creature as black people, they would behave like black people! Anyone promoting this view would be branded a bigot, and rightly so. The only difference between the statement about animals and the statement about black people is the focus of the prejudice. Just as the Nazis viewed Aryans as the master race, most people view humans as

---

[22] Singer (2002b), p. 9.

the master species. And we tend not to notice we're even doing it.[23]

Singer makes the extremely interesting and challenging point that the amount of suffering and pain caused by the tyranny of human beings over other animals (particularly in food production) far exceeds that caused by sexism, racism, or any other existing form of discrimination, and that for this reason *the animal liberation movement is the most important liberation movement in the world today.* Women and disadvantaged ethnic groups have never been farmed, killed for sport, or systematically experimented on in anything like the numbers that non-human animals have. Furthermore, unlike women and slaves, non-humans cannot talk or campaign for their own liberation, and, because they can't vote, they're not a high priority for most politicians. This further underscores the importance of the animal liberation movement.

None of the ethical conclusions we've surveyed here are logically necessary implications of evolutionary theory, and it is certainly not the case that everyone who accepts the theory accepts these ideas. The reason evolutionary theory is important is that these kinds of

---

[23] For readers diligent enough to read the footnotes, here's another example. Several years ago, the Foundation for Biomedical Research ran an ad campaign in support of medical experimentation on non-human animals. The ad featured a photograph of a group of animal rights protesters under the caption: 'Thanks to animal research, they'll be able to protest 20.8 years longer.' But imagine a parallel universe in which medical research is conducted on black people, and in which an equivalent foundation employs an equivalent argument: 'Thanks to research on black people, these white protesters will be able to protest against experimentation on black people 20.8 years longer'! Would this justify experimentation on black people? Obviously not! We would immediately reject the argument as founded on a deeply racist assumption, namely, that the costs inflicted on 'mere' black people.are justified by the benefits produced for whites. But the original argument is founded on an equivalently *speciesist* assumption: that the costs inflicted on 'mere' animals are justified by the benefits produced for us. My point is not that this assumption is necessarily wrong; perhaps there are differences between humans and animals that justify sacrificing the latter for the former (and then again, perhaps not). My point is that most people don't even *notice* the prejudiced assumption underlying the argument, let alone try to justify it.

ideas would be virtually unthinkable from a pre-Darwinian standpoint. Darwin shows us, if nothing else, that the ideas should be thinkable. His theory opens up the floor for debate on these issues, freed from the dogma that human life is infinitely valuable whereas the lives of non-human animals are utterly devoid of any value.

Conclusion

A widespread attitude, among religious believers and non-believers alike, is that even if the doctrine of human dignity is false, it is at least harmless. But it is far from harmless. It provides the intellectual foundations for a moral system that unblinkingly sanctions indifference to the suffering of healthy non-human animals but classes mercy killing as murder. The effect of the doctrine of human dignity is to increase the sum total of suffering in the world. Evolutionary theory undermines the doctrine, and in doing so, sweeps away much of traditional morality. It leaves the field open for a new morality, one that meshes better with our modern understanding of the universe and with enlightened goals such as preventing unnecessary suffering. A question that comes to a lot of people's minds, though, is whether, in the wake of Darwin's theory, *anything* can be considered morally right or wrong – in other words, whether evolutionary theory not only undermines the doctrine of human dignity but undermines morality altogether. That's the subject matter of the next chapter, which is the final chapter of the book.

FOURTEEN

Evolution and the death of right and wrong

Although the shrewdest judges of the witches and even the witches themselves were convinced of the guilt of witchery, this guilt nevertheless did not exist. This applies to all guilt.

Friedrich Nietzsche (1974), p. 216

Morality is a collective illusion of the genes. We need to believe in morality, and so, thanks to our biology, we do believe in morality. There is no foundation 'out there' beyond human nature.

Michael Ruse (1995), p. 250

Practise random kindness and senseless acts of beauty.

Bumper sticker

What's the big deal?

What's the big deal about evolutionary theory? Why do Creationists make such a fuss? In one sense, the answer is obvious: as we saw in earlier chapters, the theory contradicts a literal interpretation of Genesis and is therefore a direct threat to the Creationist worldview. This doesn't quite answer the question, though. Creationism, at least in its Young Earth form, is also inconsistent with geology and plate tectonics and the Big Bang theory and archaeology and linguistics and indeed any other area of human knowledge that traces its object of study back further than 6,000 or 10,000 years.[1] It is interesting and strange that there is such a strong anti-evolution movement, but no equivalent movement aiming to stifle geology or Big Bang cosmology.[2] The scientific revolution that really seems to get people hot under the collar is the Darwinian revolution. Our question, then, is why do Creationists worry almost exclusively about evolution?

Much of the opposition stems from a concern about how evolutionary theory might influence people's behaviour and affect their moral conduct. In his own day, Darwin was branded the most dangerous man in England. A lot of today's Creationists would consider this an understatement. Darwin's ideas, they believe, constitute the most harmful philosophy on the planet. If they became widely known, it would spell the end of human goodness. It would erode morality and hasten the demise of civilization.

Many of Darwin's contemporaries lost sleep over this possibility. Immediately after the *Descent of Man* was published in 1871, a concerned editorialist wrote in the *Edinburgh Review* that if Darwin's views on the evolution of morality were accepted, 'most earnest-minded men will be compelled to give up these motives by which they have attempted to live noble and virtuous lives, as founded on a mistake; our moral sense will turn out to be a mere developed instinct'.[3] Charles Lyell, one of the founders of modern geology, suggested that if evolution is true, 'all our morality is in vain'.[4] Another geologist, Adam Sedgwick, asserted that if evolution is true, 'morality is moonshine'.[5] The same fears are still expressed today (although not by

---

[1] See, e.g., Pennock (1999).
[2] Actually, such movements are just starting to take off now.
[3] W. B. Dawkins (1871), pp. 195–6.   [4] Cited in Lyell (1881), p. 186.
[5] Cited in Clark and Hughes (1890), p. 84. Note that Lyell and Sedgwick were not responding to Darwin's work, but to an earlier book arguing that evolution had taken place: Robert Chambers' *Vestiges of the Natural History of Creation* (1844).

intellectuals of the calibre of Lyell or Sedgwick). This is one of the main reasons that Creationists and other anti-evolutionists are terrified of evolutionary theory and want to prevent its spread. This chapter deals with the possibility that Darwinism undermines morality. The questions we'll attempt to answer include the following: do we think that certain actions are morally right because they really are morally right? Or do we think they're right only because this way of thinking led our ancestors to have more offspring than those who thought otherwise? Does the fact that our moral beliefs have an evolutionary origin imply that our moral beliefs are false? Does it imply, in other words, that 'morality is moonshine' and that people can just do whatever they like?

In addressing these questions, I need to tread carefully and spell out my position as clearly as I can. For although there is a sense in which I agree with the anti-evolutionists that evolution undermines morality, there is also an important sense in which I disagree. There are two distinct ways that evolutionary theory could undermine morality. First, it could lead people to stop following the moral codes of their society by undermining their faith in God and personal immortality, and by telling them that human beings are just animals or collections of chemicals. And, second, it could undermine the idea that there is an objective foundation to morality. The latter claim does not necessarily imply the former, and indeed my view is that the latter claim is true but the former false. In other words, evolutionary theory undermines the view that morality has an objective foundation, but it won't turn us into depraved egotists or murderers.

Can we be good without God?

We'll start with the suggestion that evolutionary theory is a toxic and corrupting influence, a worldview which leads people to flout and disregard the rules of morality. This claim is virtually axiomatic among US Creationists. Many believe that evolutionary theory is to blame for societal ills from materialism, selfishness, and promiscuity to violence and even genocide. The Creation 'Scientist' Henry Morris proclaimed that 'Evolution is the root of atheism, of communism, Nazism, behaviorism, racism, economic imperialism, militarism, libertinism, anarchism, and all manner of anti-Christian systems of belief and practice.'[6] (Note that this was in a book arguing that the story of Noah's Ark was a genuine historical event.) More generally, Creationists argue that, if widely understood, evolutionary theory will lead people to become untrustworthy and immoral, and lead human societies to fall into lawlessness and chaos. Civilization will come crumbling to the ground.

Not surprisingly, these kinds of claims are based less on data than on theoretical considerations. The most popular theoretical reason to think that Darwin's theory might derail morality is that it undermines belief in God and an afterlife. The argument can be framed as follows. Morality, almost by definition, involves going against self-interest. Our moral codes tell us that we should help others even when it couldn't possibly advantage us, and that we should refrain from lying, cheating, and stealing even when it probably would. Why would anyone go against self-interest in these ways? Well they wouldn't, goes the argument, but for God. God provides a reason to act in ways that are not in our best interests – because ultimately, thanks to God's intervention, such actions *are* in our best interests. People who believe in heaven and hell (if they do genuinely believe) think that no good deed goes unrewarded and no crime unpunished. This gives them a self-interested reason to be moral.[7] Evolutionary theory takes that reason away. Without the carrot and stick of eternal bliss v. eternal damnation, atheists would be more willing than believers to be bad. Of course, atheism would not eliminate moral behaviour completely; often enlightened self-interest dictates that we behave

---

[6] Morris (1972), p. 75.
[7] The same applies to the Eastern doctrine of karma and reincarnation.

in accordance with the moral principles of our society. Nonetheless, morality and self-interest part company all too often. There will always be occasions when one believes, rightly or wrongly, that one could get away with bending or breaking the rules. In those circumstances, why be moral? The answer that people are afraid of is that, after Darwin, there *isn't* any reason to be moral.

It's not just about heaven and hell, either. Another theoretical reason to suppose that evolution would erode morality is that it paints a rather bleak and degrading picture of who we are and where we stand in the grand scheme of things (see Part II). Within the Darwinian worldview, humans are mere evolved animals. Some critics claim that this gives people a licence to 'act like animals' – in other words, to engage in the types of promiscuous sexual behaviour that it is assumed all other species engage in. (This assumption is false, incidentally. Some do; some don't.) Furthermore, according to a number of commentators, if we are products of natural selection, rather than the handcrafted creations of a loving God, we could have no more value than inanimate objects. Human life would be worthless, and thus there would be nothing wrong with mistreating or even killing other human beings. To those who hold this view, atheistic evolutionists who are opposed to cruelty and killing (as most of them obviously are) have failed to recognize the true implications of their own belief system. They are trying to have their cake and eat it too: ditching God but holding onto ethical concepts that only make sense in a theistic context. The problem is that not everyone will do this. Some people will recognize the true implications of Darwin's theory and act accordingly. Indeed, we see evidence of this every day in the news. Many Creationists trace the (alleged) increase in violence in society over recent years to the teaching of evolution. For example, Thomas DeLay, a member of the US House of Representatives, claimed that one of the key factors in violent crimes such as the Columbine school massacre is the fact that 'our school systems teach our children that they are nothing but glorified apes who have evolutionized [sic] out of some primordial soup of mud'.[8] Claims along these lines have a long history. In 1900, an editorialist in *L'Univers* noted that:

> The spirit of peace has fled the earth because evolution has taken possession of it. The plea for peace in past years has been inspired by faith in the divine nature and the divine origin of man; men were then looked upon as children of one Father and war, therefore, was fratricide. But now that men are looked upon as children of apes, what matters it whether they are slaughtered or not?[9]

Let's ignore the unfortunate implication that it doesn't matter whether apes are slaughtered, and start sifting through these suggestions. As noted, opponents of evolution allege that Darwinism promotes selfishness, promiscuity, violence, and a general climate of immorality. There are two main responses to these allegations. First, even if the critics are right about the pernicious effects of evolutionary theory, this would not show that evolutionary theory is false. And, second, they're not right. Admittedly there may be less reason to be moral without God than with. After all, why would anyone promise heaven and hell if doing so didn't affect other people's behaviour?[10] But even if people really do have less reason to be good without God, this is no different from the fact that children have less reason to be good without Santa Claus. There may still be *enough* reason to be good without God. There may even be better reasons to be good.[11]

In any case, no matter how persuasive the theoretical grounds for thinking that evolutionary theory causes moral decline, the evidence simply does not stack up with this view. Consider the claim that teaching evolution increases levels of violence in society. It turns out

---

[8] DeLay and Dawson (1999).   [9] Cited in Bryan (1922), p. 124.

[10] This need not be taken as a rhetorical question. People might promise heaven and hell because they *believe* it will influence others' behaviour when in fact it tends not to. I'll say more about this soon.

[11] On goodness without God, see Buckman (2002); Nielsen (1973).

that violence is actually *less* common in parts of the world where evolutionary theory is widely taught than in parts where it's not.[12] The United States is the most religious country in the Western world – the developed country with the greatest resistance to evolutionary theory – but it's also the Western country with by far the most violent crime. Furthermore, within the US, the states with the most fundamentalist religion and thus the least evolution – i.e., the Southern states – are also the most violent. This is hard to square with the idea that the teaching of evolution causes an increase in violence. If anything, it has the reverse effect!

OK, responds the Creationist, maybe evolutionary theory doesn't encourage gross violations of the moral code, such as violence or murder. But what about less serious offences? If people cease to believe in God, won't they become more selfish and antisocial in their day-to-day lives? An initial point to make is that this idea only makes sense on the assumption that people are naturally this way inclined to begin with. Only someone who thinks that people's natural inclinations are selfish and antisocial would assume that, without God, people would behave in a selfish or antisocial manner. In other words, this view betrays a low opinion of human nature. Now this doesn't mean it's wrong, but as it happens, it does appear to be wrong. It involves taking an overly dim view of human nature. A *somewhat* dim view of human nature accords better with the facts. As we saw in Chapter 11, although we've got a selfish streak, we're not completely selfish; we've got a definite altruistic streak as well. We have a natural capacity to empathize with other people (or at any rate, with members of groups to which we belong), and we routinely help people out in times of need. Our moral sense is part of human nature. It predates Jesus and Moses and Confucius and all the other moral teachers of the ages. Indeed, it can be traced back into our pre-human past, and hints of it can be found in other species. This last point raises

[12] Isaac (2005).

another issue: plenty of non-human animals live in cohesive social groups, despite their lack of belief in either God or the immortality of the soul. Why should we think that our species is any different?

A sophisticated Theistic Evolutionist might have an answer to this. Unlike most animals, modern humans live in conditions that in many ways are unlike those we're adapted for. Human nature is tailored primarily for the small, face-to-face hunter-gatherer societies in which such a significant part of our evolution took place, and in today's anonymous, large-scale societies, human nature may no longer be enough. To live together peacefully in modern conditions, we may need something extra, some kind of intellectual prosthesis to constrain and control our natures, and to augment our natural goodness. This sounds to me like a reasonable suggestion. But even if people living in mass societies really do need social institutions to boost their natural levels of goodness, what makes us think these need to be *religious* institutions? What makes us think they must be founded on a view of the universe that is patently false? Is the only way to deal with, say, bullying in schools to tell the bullies that an all-powerful disembodied mind is watching them and will magically punish any transgressions? Clearly not. Educators have devised all sorts of anti-bullying interventions, and have empirically demonstrated their worth. We can teach children morality without tying it to an implausible metaphysic. In fact, it's possibly safer and more effective to do so. Tying morality to religion is a little like transporting a precious cargo on a sinking ship. What happens when the child grows up and starts doubting the factual claims of the religion? The cargo may be lost with the ship.[13]

Up until now, we've been asking whether atheism makes you bad. However, it's only fair to ask a follow-up question as well: does religion make you good? Most people – even those who are not religious themselves – assume that it does. Unfortunately, the case

---

[13] On bringing up ethical children without God, see Dale McGowan's (2007) collection of essays.

is not nearly as strong as people think. Earlier we discussed the religious sanctions of heaven and hell. Now we'll ask just how effective these sanctions really are. There are at least three reasons to question their effectiveness. First, if God and heaven and hell were enough to make people good, institutions such as the police and prisons would only be needed for atheists. This is not the case. In fact, it is sometimes suggested that atheists are *less* likely to be incarcerated than believers, although I've been unable to find any properly controlled research to back up this claim. Second, even people who supposedly believe in heaven and hell are often less influenced by these distant possibilities than they are by more imme-diate carrots and sticks, such as social approval and imprisonment – things that are also concerns for atheists. And, third, many of the people we consider the most evil sincerely believe that what they're doing is right, and the threat of hell is not going to stop these individuals. Instead, the promise of heaven may inspire them to do things the rest of us consider profoundly wrong. Case in point: the Islamic terrorists who hijacked airplanes and slammed them into the World Trade Center and the Pentagon on 11 September 2001 sincerely believed that their act was noble and right, and that it would earn them and their loved ones entrance into a particularly pleasant part of paradise. And we all know about the terrible things done in the name of religion in earlier centuries: the Crusades, the Spanish Inquisition, the burning of so-called witches, etc.[14] Throughout history, and even today, people have persecuted and killed one another on the basis of mere superstition.

Believers often write these things off as aberrations and deviations from the true spirit and true teachings of religion.[15] Of course, those perpetrating the acts would say the same thing about these believers, and there is no objective way to determine who's right. But the case

[14] Ellerbe (1995).
[15] Ward (2006).

against religious morality isn't limited to the exceptional cases. When we look at large-scale surveys of everyday believers, we find that in many ways atheists are actually more moral than believers. On average, they are less prejudiced, less racist, and less homophobic; more tolerant and compassionate; and more law-abiding.[16] Admittedly, whether this means that atheists are more moral depends on your personal convictions; if you think homophobia is a virtue, for instance, then you'd have to conclude that a greater number of religious than non-religious people possess this particular virtue. Nonetheless, a convincing case can be made that non-religious moral codes are often superior to those traditionally linked with theism. Consider Peter Singer's Top Three ethical recommendations: do something for the poor of the world; do something for non-human animals; and do something for the environment. This is the ethic of an atheist, a man who accepts that life evolved and has no ultimate meaning or purpose. To my mind, it is vastly superior to moral systems emphasizing trivial issues (or non-issues) such as premarital sex, blasphemy, and the like. Morality is not just about deciding what's right and wrong, good or bad. It involves getting your moral priorities straight.

To be fair, plenty of religious people do have their moral priorities straight. Many don't, however. For example, if we judge them by where they put their efforts and what they spend their time worrying about, we would have to conclude that many conservative Christians consider premarital sex more morally problematic than human rights abuses, and homosexuality more morally problematic than the mistreatment of animals or the environment. In other words, they consider it more important to eliminate the *pleasures* of certain groups of individuals than it is to eliminate the *suffering* of certain other groups. The fact of the matter is, though, that this moral priority list is easier to justify in terms of scripture than are the priorities of the liberal

---

[16] See Batson *et al.* (1993); Beit-Hallahmi (2007); Paul (2005); Zuckerman (2008).

religionists.[17] The Bible and the Koran are quite unambiguous in their denunciation of homosexuality,[18] for example, but provide little unambiguous guidance on issues such as animal welfare. If the stories and moral teachings of these ancient texts provide an authentic guide to morality, then modern people – including most modern theists – have slipped dangerously away from the path of righteousness, with our heinous devotion to ideals such as equality, the respectful treatment of women, and tolerance of people who happen to be sexually attracted to members of their own sex. To the degree that religious people have their moral priorities in good order, this is often *in spite* of their religion rather than because of it. One could even argue that they are adopting a secular code of morality but then crediting it to their religion – a religion that actually teaches something quite different. Bearing all this in mind, we're left with a question: if religion doesn't make people good, why do people think that it does? Simple: because religion *teaches* that it makes people good. It's part of its sales pitch. But it's also quite possibly untrue.

That said, we do have to acknowledge that both religionists and atheists have done good things and bad things, and that to a large extent the buck must stop with human nature, rather than with either of these belief systems. It's fair to say that, with or without religion, most people are good. They're not *great*, but they are good.

The myth of morality

As mentioned, there are two ways that evolutionary theory might sound the death knell for morality. We've just discussed the first: the theory could undermine morality by undermining theistic belief and human dignity, thereby encouraging selfishness, violence, and bad

---

[17] Harris (2004).

[18] For biblical denunciations of homosexuality, see, e.g., Leviticus 18:22, 20:13; Romans 1:26–7. For Koranic denunciations, see, e.g., Sura 7:80–1, 26:165–6, 27:54–5, 29:28–9.

behaviour in general. We found this claim wanting. The second way that Darwin's theory could undermine morality is that it could undermine the idea that there are objective moral truths – truths that exist independently of human minds, emotions, and conventions. In the remaining pages of this book, I'm going to argue that evolutionary theory does indeed undermine this idea, and that morality is, in some sense, a human invention (or, more precisely, a joint project of human beings and natural selection). In other words, in the final analysis, nothing is right and nothing is wrong. This perspective is quite counterintuitive to most people (myself included). Like a lot of topics in philosophy, though, the opposite position (*moral realism*) is counterintuitive as well. Moral realism is the view that there are objective facts concerning what is right and wrong, and that true moral beliefs correspond somehow to these moral facts of the universe. To most people, this way of talking sounds a bit odd. However, it is more closely aligned to our intuitive view than is moral *anti-realism*: the idea that there is no reality to our moral judgments. People do tend to feel, for instance, that the statement 'murder is wrong' is true no matter what anyone thinks and, more generally, that our moral principles are things we know, rather than just things we've made up. So, as peculiar as it may sound to talk of objective moral truths, people do act as though they believe in them.

One way to start tackling the issue of the objectivity of morality is to revisit a question we posed in Chapter 11: where does our knowledge of moral truths come from? Various answers have been given, some of which are reasonable but many of which set off alarm bells in heads that are well screwed on. Some suggest that God planted innate knowledge of moral truths in our minds. Others suggest that human beings have something called a conscience: an inner voice through which God reveals what is right or wrong. (Strangely enough, the rights and wrongs that God communicates tend to be almost identical to those espoused by people's parents and peers, but very different from those found in other cultures.) A related suggestion is that we possess a mysterious faculty of moral intuition through which we directly perceive objective moral truths, in the same way that, with the visual sense, we directly perceive objective facts about the physical world.[19] Another important answer is linked to the notion that science and religion have distinct and non-overlapping domains. The idea is that science is limited to providing empirical knowledge, whereas religion's role is to provide knowledge of moral truths, through divinely revealed scriptures or the pronouncements of religious leaders. The position to be explored here is that any and all such suggestions must be rejected in the light of evolutionary theory. The argument is *not* with the claim that science is powerless to supply us with moral knowledge; no one imagines that there could ever be an experimental procedure that would directly detect the rightness or wrongness of any action or intention. The argument is that, rather than religion or intuition supplying us with knowledge of moral truths, knowledge of moral truths is not possible, *for there are no moral truths*.

This view should not be confused with *moral relativism*, which is, in effect, the view that all moral beliefs are true, at least within the cultures in which they are held. The position under discussion is *moral nihilism*, the view that all moral beliefs are *false*, because all are equally groundless. According to moral nihilists,[20] we naturally believe that moral statements are objectively true, rather than things we invent. But we're wrong about this. As Nietzsche put it: "'This – is just *my* way – where is yours?' Thus I answered those who asked of me "the way". For *the* way – does not exist!'[21]

The most plausible alternative to moral nihilism is *non-cognitivism*. According to non-cognitivists, moral nihilists are wrong to think that all moral beliefs are false. *No* moral belief is false – but nor is any moral belief true. Moral beliefs are neither true nor false, says

---

[19] Moore (1903).
[20] E.g., Joyce (2001); Mackie (1977); Williams (1972).
[21] Nietzsche (2005), p. 169.

the non-cognitivist, because moral statements were never intended to report facts about the world. Some non-cognitivists suggest that ethical statements reflect the preferences or attitudes of the speaker. To say 'murder is wrong', for example, is really to say 'I don't like murder.' Another view is that moral statements are veiled commands.[22] Thus the statement 'killing is wrong' really means 'don't kill'. Either way, moral statements are neither true nor false, any more than statements such as 'hooray for chocolate' or 'don't steal my chocolate' are true or false. They're not the kind of statements that *can* be true or false.

The ethical nihilist rejects non-cognitivism for several reasons.[23] First, if we meant something like 'hooray for chocolate' or 'don't steal my chocolate', why wouldn't we just *say* that, rather than using the form of language we use when asserting facts? Second, the fact that people disagree over moral matters implies that different moral principles clash with each other, which in turn implies that – you guessed it – they're intended to express facts. If I say 'crème brûlée is delicious' and you say 'no, it's horrible', there is no logical contradiction to be cleared up; we just have different tastes. But if I think abortion is permissible and you think it's murder, we plainly disagree. This suggests that people see moral statements as the type of statements that can be true or false.[24] Certainly, we can *infer* from the fact that people say 'X is wrong' that they disapprove of X and that they probably don't want anyone to do X. But that is not what they *mean* by 'X is wrong'. When people make moral pronouncements, they are 'telling it like it is' – or rather that's what they think they're doing. The ethical nihilist denies that what they're saying is correct. Let's see why.

[22] Hare (1989).

[23] The arguments in this paragraph are based on Joyce (2001).

[24] The secularist philosopher Austin Dacey (2008) argues that moral disagreement implies that morals are objectively true. However, it implies only that we view them as such.

Atheism and the death of morality

There are two ways that an evolutionary perspective might support ethical nihilism and undermine the existence of objective moral truths. The first is by undermining the existence of God. (Remember we're talking now about how atheism might challenge the objectivity of morality, not about how atheism might lead people to behave immorally.) A lot of people think that if there is no God, there can be no real right or wrong. In Dostoyevsky's book, *The Brothers Karamazov*, the character Ivan Karamazov famously declares that, 'If God does not exist, then everything is permitted.' This captures the view that morality stems ultimately from God and is dependent on God. The formal name for this view is *Divine Command Theory*. According to proponents of Divine Command Theory, moral principles are commands and, just as you cannot have design without a designer, you cannot have commands without a commander. *Sans* God, morality becomes arbitrary and ultimately ungrounded. Anyone could claim that any standard is moral, and there would be no objective yardstick to prove or disprove it. Any moral claim would be as good as any other (or, according to taste, as bad as any other). If this view – that God is the source and justification of morality – is accurate, then to the degree that evolutionary theory is inconsistent with the existence of God, the theory eliminates morality's source and justification.

People often take it for granted that, without God, there can be no objective morality, and this is an issue that is debated routinely in discussions about the existence of God. A discouraging fact about these debates is that, as far as most philosophers are concerned, the issue was decisively resolved more than 2,000 years ago. One of the oldest criticisms of the idea that morality must be dependent on God – and still one of the best – is known as the *Euthyphro dilemma*, and was outlined by Plato. The question Plato asked (through his mouthpiece, Socrates) was: 'Is the pious loved by the gods because it is pious, or is it pious because it is loved by the gods?' Translated into monotheistic terms, the question becomes: 'Are the things that God commands good because God commands them, or does God command them because they are good?' This little question, which sounds so innocuous at first, opens up a great big can of worms for Divine Command Theory.

Imagine, first, that the theist answers that a given act is morally right purely and solely because God commands it. The most obvious difficulty with this position is that we would then have to accept that, were God to command us to torture children or commit murder for sport, this behaviour would be good and right, just because God said so. This is completely contrary to our moral intuitions (which supposedly come from God in the first place). The theist might object that God would never command such terrible things; God is good and would therefore only command what is good. This objection doesn't work, though, because it tacitly assumes that there is some standard of goodness independent of God's decrees. If the good is whatever God commands, then to say 'God commands what is good' is to say only that 'God commands what God commands'. Oops! Another disconcerting implication is that if the good is just whatever God says it is, morality has an arbitrary foundation: the whims of God. Some theists are willing to bite the bullet and accept that our moral rules are ultimately arbitrary. But most are not. After all, one of the initial complaints against *secular* morality was that it is arbitrary.

Having weighed up these arguments, most theists change their tune and renounce the view that good acts are good only because God commands them. Instead, they decide that God commands certain acts *because* those acts are good. What are the consequences of this move? Well it does avoid the problems we've considered so far. But there's a price to be paid: it makes morality independent of God. God is no longer the ultimate author or foundation of morality. He commands certain acts because they are good, and they would be good even if he did not command them – or even if he did not exist.

The important point is this: as soon as we concede that objective morality could exist independently of God, we must also concede that atheism does not automatically imply the end of morality.[25] Thus, if evolutionary theory undermines objective moral truths, this is not because it undermines belief in God.

Nihilism and the evolution of morality

There is another way that evolutionary theory encourages scepticism about objective moral truths. As we saw in Chapter 11, promising evolutionary explanations have been proposed for some of our most basic moral inclinations and feelings. Many of these things are deep-seated preferences crafted by natural selection. According to some philosophers, such as Michael Ruse and Richard Joyce,[26] these results reveal that our moral beliefs are illusions, held not because they are true but because they are useful in regulating the social life of a highly social animal. To the degree that morality is a product of natural selection, it is an illusion foisted upon us to propagate the genes disposing us to be moral.

There are a number of ways that this argument can be fleshed out. One is to appeal to the law of parsimony (Occam's razor). Imagine two universes, one in which there are objective moral values and one in which there are not. The laws and the initial arrangement of matter in each universe are absolutely identical. Would the evolution of human morality differ across these two universes? It would not. Altruism towards kin would be selected in both; reciprocity would be selected in both; an aversion to incest would be selected in both. In short, the same selective forces, discussed in Chapter 11, would produce the same moral psychology in both universes. Why, then, should we assume the existence of objective moral values? Occam's trusty razor tells us that, if our moral judgments would be the same

---

[25] Brink (2007).
[26] Ruse (1995); Joyce (2001).

regardless of whether there are objective moral truths or not, we should assume that there are no objective moral truths. In the words of Michael Ruse and E. O. Wilson:

> The evolutionary explanation makes the objective morality redundant, for even if external ethical premises did not exist, we would go on thinking about right and wrong in the way that we do. And surely, redundancy is the last predicate that an objective morality can possess.[27]

The philosopher Richard Joyce put forward another argument against the objectivity of moral truths, an argument with a similar flavour but which did not rely on an appeal to parsimony. According to Joyce, the important point is not that moral objectivity is unparsimonious but that it is *unjustified*.

> The innateness of moral judgements undermines these judgements being true for the simple reason that if we have evolved to make these judgements irrespective of their being true, then one could not hold that the judgements are *justified*. And if they are unjustified, then although they *could* be true, their truth is in doubt.[28]

Of course, we concluded earlier that morality is not actually innate, but is a human invention informed by preferences and motivations that *are* innate. It is a product of both biological and cultural evolution. The general point remains, though. With a naturalistic account of morality in hand, we no longer need to posit objective moral values, or commandments issued from a divine lawgiver, in order to explain our moral judgments and intuitions. Our moral beliefs are the product of a process that does not presuppose the truth of those beliefs. As soon as we realise this, we are no longer justified in holding those beliefs, unless we can find independent grounds for doing so. The mere fact that we hold them – even the fact that they seem inescapably compelling to us – can no longer be considered an adequate reason to accept them as true.

[27] Ruse and Wilson (2006), p. 566.
[28] Joyce (2001), p. 159.

There are a number of standard objections to the evolutionary argument against moral truths. The most common is that it commits a fallacy of reasoning known as the *genetic fallacy*. The genetic fallacy has nothing to do with genes or heredity. It refers to the origin (or genesis) of a belief. The idea is this: the fact that we can explain why someone holds a particular belief does not imply that that belief is false. If I knew your life story in intimate detail, I could probably explain how, during the course of your childhood, you acquired the belief that the earth orbits the sun. In doing so, however, I would not have proved that the earth does *not* orbit the sun. Similarly, even if evolutionary theorists can explain how we ended up holding the moral beliefs that we do, they have not proved that these beliefs are false. It could still be the case that there are objective moral truths. Put simply, the truth of an evolutionary explanation for moral beliefs does not *entail* the conclusion that there are no objective moral truths.

Now I'm quite willing to concede this point. But it's not much of a concession. After all, it is equally true that an evolutionary explanation for, say, food preferences does not entail the conclusion that there are no objective truths about which foods are delicious and which unpalatable. Fair enough, but why assume that there *are* objective culinary truths? Unless positive reasons can be adduced for believing in objective moral truths, the argument amounts to the assertion that 'there *could* be objective moral truths; it's not logically impossible'. If anyone were willing to accept this as an adequate reason to accept the existence of moral truths, then, to be consistent, that person would also have to accept the existence of Santa Claus – after all, it's not logically impossible that Santa exists. And that's just the start. This person would have to accept the existence of *all* logically possible entities or states of affairs. So, if this is all that can be said in support of objective moral truths, we could be forgiven for remaining unconvinced. There's another problem as well: although it is logically possible that there are moral truths, it is also logically possible that there are not. These two possibilities cancel each other out, and we're left back where we started. When the best we can do is appeal to the fact that a premise is logically possible, we should acknowledge that we have no good reason to believe the premise in question. We are clutching at straws.

Perhaps, though, we're dismissing the argument too quickly. There is a related but higher calibre objection to the proposition that evolutionary theory undermines morality.[29] Think about other evolved competencies – the capacity for vision, for example. We know this capacity has an evolutionary origin. Does that imply that nothing we see is real? Of course not! Or take the ability to reason about the motion of objects (folk physics) or about minds (folk psychology). Again, a good case has been made that there is an evolved contribution to these abilities.[30] Does this imply that every statement made by a physicist or psychologist is false? No! Not even the latter claim is plausible. Why, then, would we assume that, if morality has an evolutionary origin, our moral beliefs must be false? To put it another way, if the argument works for our moral beliefs, it must also work for our visual beliefs, our beliefs about objects, and our beliefs about other minds. But it would be ludicrous to conclude that all our beliefs in these other domains are false, and thus we can safely ditch the argument as applied to our moral beliefs.

Or can we? The above objection ignores an important distinction between our moral beliefs and our beliefs about the physical world and other minds. The evolutionary function of vision, folk physics, and folk psychology relates to the accuracy of our understanding of the world. The evolutionary function of our moral beliefs, on the other hand, is unrelated to their objective truth or falsity. It revolves *solely* around the effects that these beliefs have on the way we treat one another. From an evolutionary perspective, it would not matter how much or how little our moral beliefs correspond to any objective

<hr>

[29] See, e.g., Dacey (2008).
[30] Baron-Cohen (1995); summarized in Pinker (1997).

moral truths, even if such truths existed. We can ask this question: if the objective facts were different, would natural selection still favour the same tendency of belief? If the answer is 'no', we have some reason to think that the belief in question is at least approximately true. On the other hand, if the answer is 'yes', then an evolutionary perspective offers us no reason whatsoever to think that the belief is true. In principle, it might be, but the mere fact that we have an evolved tendency to think it true is no proof that it actually is.[31] Some beliefs are adaptive because they correspond to objective facts about the world. Moral beliefs are not of this type. They would be adaptive *whether or not* they corresponded to objective moral truths. We therefore have no reason to think that such truths exist.

How might the moral realist respond to this? One option would be to reformulate the argument, but to focus not on our perceptual faculties, but instead on our mathematical competency. Our basic facility with numbers plausibly has an evolutionary origin.[32] This would not, however, lead us to conclude that mathematical statements, such as $1 + 1 = 2$, are false. Nor should it in the case of ethical statements. An advantage of this argument over the earlier version is that mathematics may provide a good model for understanding how ethical statements could be objectively true, despite the fact that they don't correspond to empirical facts in the way that perceptual beliefs do. Mathematical statements don't either, but we're perfectly happy to accept that *they* can be objectively true. Of course, we might question whether this is a reasonable view to hold; maybe it's a mistake to think we can assign truth values to mathematical statements. Some philosophers do take this position, but I suspect that they're wrong, for a number of reasons. First, progress in mathematics has the character of discovery rather than invention. This suggests that mathematics involves uncovering truths, rather than just making stuff up. Furthermore, different cultures

---

[31] Cf. Stewart-Williams (2005).

[32] Butterworth (1999); Dehaene (1997); Geary (1995).

converge on the same basic truths of mathematics, which suggests again that mathematical truths really are true. They're abstract or conceptual truths, certainly, but they're truths nonetheless – truths that would presumably hold in any possible universe.

Some suggest that ethics is the same. Just as different cultures settle on the same mathematical beliefs, so they tend to settle on the same moral imperatives, such as the Golden Rule. Perhaps *any* intelligent social species anywhere in the universe would eventually discover the Golden Rule and other foundational ethical principles. Perhaps true moral beliefs are abstract or 'Platonic' truths about the best and most effective ways to organize societies of intelligent beings. Perhaps ethical behaviour is what Daniel Dennett calls a Good Trick – a design so useful that natural selection is bound to 'discover' it again and again, whether through biological or cultural evolution.[33] Based on considerations such as these, we may be tempted to agree with Robert Nozick's summation of the issue:

> We cannot yet specify, even roughly, how we know ethical truths. However, our inability thus far to develop an adequate epistemology for mathematics, to specify how we can know *those* truths, has not pushed us to conclude that there are no such truths. Why should things be different with ethics?[34]

This is a good question, but I think I have a good answer. Again, the key is a consideration of the purposes for which different aspects of our psychology evolved. Clearly, we would not want to argue that the evolutionary origin of basic mathematical competency implies that mathematical statements such as $1 + 1 = 2$ are false. However, there's an important difference between mathematical statements and moral judgments. The evolutionary story about the origins of mathematical competency relies on the assumption that mathematical statements are in some sense true and applicable to the real world, despite their

[33] Dennett (1995).
[34] Nozick (1981), p. 343.

abstract character. Ruse provided a simple example of how an understanding of number could contribute to evolutionary success: 'Two tigers were seen going into the cave. Only one came out. Is the cave safe?'[35] Admittedly, it's difficult to say what in the world mathematical statements correspond to that makes them true. However, the 'unreasonable effectiveness of mathematics in science'[36] suggests that the abstract truths of mathematics are somehow related to reality in a profound way.[37] Ethics is different. Our story about the evolution of morality need make no reference whatsoever to the truth of our moral beliefs. All that is required is that we act in a way that, in the past, proved useful to any genes contributing to the development of these beliefs. This is why the evolutionary argument applies to ethical beliefs but not to mathematical ones.

There's another important difference between ethical and mathematical beliefs. Intelligent life forms on any planet would presumably come to the same mathematical understandings. Why, though, should we assume this of ethics? In as much as our ethical intuitions have an evolutionary origin, they may fit our species to its evolved lifestyle in a manner that is relatively unique. If it seems that the ways we have hit on are the only possible ways, perhaps this simply represents a failure of imagination. Cannibalism and incest are moral no-no's for most humans, but members of other species have evolved to engage in these activities with relish. Who are we to doubt that an animal with this sort of alternative lifestyle could ever evolve into a rational, moral animal without relinquishing those tendencies? Darwin wrote this on the subject:

---

[35] Ruse (1986), p. 162.   [36] Wigner (1960).

[37] The nature of mathematics and its relationship to the physical world is actually a deep and extremely difficult philosophical question. However, it is one that people often fail to appreciate. The mathematician R. W. Hamming (1980) reported that: 'I have tried with little success to get some of my friends to understand my amazement that the abstraction of integers for counting is both possible and useful' (p. 81).

I do not wish to maintain that any strictly social animal, if its intellectual faculties were to become as active and as highly developed as in man, would acquire exactly the same moral sense as ours. In the same manner as various animals have some sense of beauty, though they admire widely different objects, so they might have a sense of right and wrong, though led by it to follow widely different lines of conduct. If, for instance, to take an extreme case, men were reared under precisely the same conditions as hive-bees, there can hardly be a doubt that our unmarried females would, like the worker-bees, think it a sacred duty to kill their brothers, and mothers would strive to kill their fertile daughters; and no one would think of interfering. Nevertheless, the bee, or any other social animal, would gain in our supposed case, as it appears to me, some feeling of right or wrong, or a conscience.[38]

And if we're looking for disagreement in matters of ethics, it's not as if we have to look to other species. For anyone who understands the meaning of the words, it is true that 1 + 1 = 2. But people constantly disagree about what's right and wrong. Certainly, there are some striking commonalities across the moral codes of different cultures, and perhaps there really are abstract, Platonic truths about how to organize societies of intelligent beings. An example might be: 'Unless there are major asymmetries in power, reciprocity works well because everyone benefits and no one gets too disgruntled.' However, to say there are abstract truths of this nature is very different from saying that there are objective moral truths. We evolved to think 'reciprocity is good' because it coordinated us with the abstract truth that reciprocity is useful for those who engage in it, *not* because it coordinated us with the abstract truth that reciprocity is good. Likewise, it might be objectively true that '*if* you want to increase the chances of people treating you well or of society running smoothly, you should treat others as you would have them treat you'. But the Golden Rule, stated without that kind of qualification, is not objectively true. Technically, it is false.

[38] Darwin (1871), pp. 122–3.

One last point: starting with Kant, a number of thinkers have suggested that morality can be conjured out of nothing from pure logic, in the same way that mathematics can. And it is true that the application of reason in the domain of ethics can lead us to new conclusions – conclusions we did not anticipate and which take us by surprise.[39] However, this is the case only once we've settled on one or more foundational ethical principles (see Chapter 12). To illustrate, imagine we decide that a universe with less suffering is better than one with more, and that we therefore adopt the ethical principle that we should aim to reduce suffering. We can then combine this ethical principle with facts about the world, and deduce new and more specific ethical directives. For instance, we can deduce the conclusion that, in as much as non-human animals suffer as we do, we should aim to reduce their suffering as well as our own. Reason would compel us to adopt this ethical conclusion – but only after we've accepted the initial ethical premise. What reason cannot do is compel us to accept this or any other principle as the foundation upon which we erect our moral system. We've got to start somewhere, without any justification. Our starting point is always a disposition, a feeling, an arbitrary assumption that something is desirable morally. And there is no ultimate justification for this. Our fundamental moral commitments cannot be derived from pure logic, or from any admixture of logic and fact.

Not surprisingly, the conclusion that there are no moral truths is easier to accept when we're dealing with a moral claim such as 'premarital sex is wrong' than when we're dealing with a claim that everyone agrees with and feels strongly about, such as 'murder is wrong'. But my argument is not that we should accept that murder is morally permissible. My argument is that there is no objective truth regarding the permissibility of murder. It is not contrary to reason to insist that murder should be treated as a heinous crime, even though there is no objective truth concerning its heinousness. This is an important point, and the next section is devoted to it.

[39] Singer (1981).

Evolutionary psychology and utilitarianism

Believe it or not, none of what I've said above implies that morality must be consigned to the scrapheap of intellectual history. All it implies is that it has no ultimate justification of the kind that people often crave. We can have morality, but it must be morality without metaphysical foundations.[40] Once we accept this, though, we may find that evolutionary principles help us erect and justify a moral system fitting this job description. In this penultimate section of the book, I'm going to argue that evolutionary psychology provides the groundwork for an argument for utilitarianism.

Utilitarianism is based on the idea that what matters in determining whether an act is good or bad is the effect that the act will have on all affected by it. Usually this is judged in terms of the production of happiness or pleasure, the satisfaction of preferences, or the alleviation of suffering. Utilitarianism's main rival is *deontological ethics*, according to which certain acts are right or wrong independently of their effects. Deontological ethics is often justified by an appeal to our moral intuitions, e.g., we just *know* that lying and stealing are wrong, and they'd be wrong even on those occasions when they happen to have desirable effects. If, however, our moral intuitions came about through a Darwinian process, then whether we acknowledge it or not, when we follow these intuitions we are acting for certain ends. To the extent that the moral sense is shaped by gene selection, we are acting for the good of the genes contributing to the development of the moral sense; to the extent that morality is shaped by memetic selection (i.e., the survival of the fittest 'memes' or units of culture),[41] we are acting for the good of our moral memes; to the extent that morality is shaped by the interests of elites, we are acting in the interests of elites, etc. We cannot avoid thinking about the effects that our moral actions have, because our moral actions exist *as a result*

[40] Joyce (2001); Maisel (2009).
[41] Blackmore (1999); Dawkins (1989); Dennett (1995, 2006).

of the effects that they have. Consequently, it would be morally negligent not to make ourselves aware of these effects, so that we can decide for ourselves whether we deem them appropriate. We should make explicit the ends to which our moral beliefs are presently directed, and if we decide we prefer other ends, we can modify our moral beliefs accordingly.

Here's another way to think about it. Philosophers distinguish between hypothetical imperatives and categorical imperatives. An example of a hypothetical imperative is: '*If* you want to avoid going to prison, *then* you shouldn't murder anyone.' An example of a categorical imperative is: 'You shouldn't murder anyone.' Categorical imperatives dispense with the 'if-clause' clause; they are out-and-out statements about what you should or shouldn't do. You should not murder anyone even if you would be perfectly happy to go to prison; you should not murder anyone, full stop. Moral statements are categorical imperatives. At any rate, that is the traditional way of construing them. However, there are hypothetical imperatives implicit in our categorical moral statements. To the extent that morality is a product of biological evolution, the hypothetical imperative implicit in any moral rule is 'if you want to propagate your genes, do X'. To the extent that morality is a product of memetic evolution, the hypothetical imperative is 'if you want to propagate this moral meme, do Y'. Get the idea? You'll often hear it said that the very essence of moral judgments is that they are *not* based on an evaluation of the costs and benefits to the actor. But the fact of the matter is that the process of natural selection has done the cost-benefit analysis for us, metaphorically speaking. To follow one's duty without concern for its effect is *not* to follow some objective and disinterested principle of goodness. It is to follow an unspoken hypothetical imperative set in place by an amoral natural process. Morality is about consequences, whether we're aware of those consequences or not. As soon as we recognize this, we can start deciding which consequences are most important to us, rather than simply accepting natural selection's 'decisions' by default.

Of course, our decisions are influenced by our evolved nature, and thus our ethical codes can never completely transcend or escape our evolutionary origins (or, for that matter, the contribution of our social conditioning). How, then, can we show that the utilitarian's goals are good – in other words, that utilitarianism is a good moral theory and one we should adopt? Well, when it really comes down to it, we cannot. The utilitarian value system will be accepted by those who, on reflection, decide that the amount of joy and suffering in the world is more important than an unquestioned allegiance to ethical principles that emerge from the competing forces of biological and cultural evolution. Note, by the way, that making an ethical commitment of this nature is perfectly consistent with an acceptance of the godless, naturalistic view of the universe that we've established in this book. Even in a pointless universe, pointless happiness and pleasures are surely preferable to pointless suffering. Nonetheless, we shouldn't dodge the hard conclusions, and the bottom line is that the utilitarian's goals – avoiding suffering and increasing happiness – are not ultimately justifiable. They just happen to be to my taste and perhaps to yours as well. I don't believe it is possible to justify moral commitments beyond this, so I will not attempt to do so.

The last word

Does this provide a satisfactory foundation for ethics? It depends who you ask. It's good enough for me, but for a lot of people it's not. The fact is, though, that it provides a stronger foundation than the idea that God laid down a set of moral rules for the world, for the simple reason that the latter is not true. To some people, the view I've outlined in these pages represents the end of morality. To me, it represents morality stripped of superstition. Ethical rules do not reflect objective moral truths any more than social conventions (such as going to the back of the queue) reflect objective civic truths. The rules of queuing are merely conventions governing situations in which there are potential conflicts of interest. Morality is the same. Think of it as being like the rules of chess. It is no more a fundamental law of the universe that stealing is wrong – or even that murder is wrong – than it is that bishops can only move along unobstructed diagonal pathways on the chessboard. Nonetheless, for the same reason that we adhere to and enforce the rules of chess, we adhere to and enforce the rules of morality: not because they reflect timeless truths residing in the mind of God or in some Platonic realm, but because they bring certain desirable effects. Put simply, they make the game more enjoyable for those involved. There is no deeper justification for morality.

Some may find these conclusions frightening, and perhaps that's an appropriate reaction. But then again, maybe it's not. For it is certainly possible to frame an ethic consistent with the Darwinian view of the world. Such an ethic might emphasize the virtue of being honest enough and courageous enough to acknowledge unflinchingly that there is probably no God, no afterlife, and no soul; that there is no objective basis to morality or higher purpose behind our suffering; that we are insignificant in a vast and impersonal cosmos; that existence is ultimately without purpose or meaning; and that the effects of our actions will ultimately fade away without trace. It is admirable to acknowledge these uncongenial truths, yet to struggle on as if life *were* meaningful and strive to make the world a better place anyway, without promise of eternal reward or hope of ultimate victory. Of course, nothing can be said to argue that people are morally obliged to accept this ethic, for to do so would be inconsistent with the ideas that inspired it in the first place. It is an ethic that will be adopted – if at all – by those who find a certain stark beauty in kindness without reward, joy without purpose, and progress without lasting achievement.

Suggestions for further reading

EVOLUTIONARY THEORY

Darwin, C. (1859). *On the origin of the species by means of natural selection.* London: Murray.

Dawkins, R. (1989). *The selfish gene* (2nd edn). New York: Oxford University Press.

EVOLUTION AND RELIGION

Brockman, J. (ed.) (2006). *Intelligent thought: Science versus the intelligent design movement.* New York: Vintage.

Dawkins, R. (2006). *The God delusion.* New York: Houghton Mifflin.

Kurtz, P. (ed.) (2003). *Science and religion: Are they compatible?* New York: Prometheus.

Ruse, M. (2001). *Can a Darwinian be a Christian? The relationship between science and religion.* Cambridge: Cambridge University Press.

Shermer, M. (2006). *Why Darwin matters: The case against intelligent design.* New York: Times.

EVOLUTION AND PHILOSOPHICAL ANTHROPOLOGY

Dennett, D. C. (1991). *Consciousness explained.* New York: Little, Brown.

Pinker, S. (1997). *How the mind works.* London: Penguin.

Radcliffe Richards, J. (2000). *Human nature after Darwin*. London: Routledge.

Williams, G. C. (1966). *Adaptation and natural selection: A critique of some current evolutionary thought*. Princeton, NJ: Princeton University Press.

Wright, R. (1994). *The moral animal: The new science of evolutionary psychology*. New York: Pantheon.

EVOLUTION AND ETHICS

Joyce, R. (2001). *The myth of morality*. Cambridge: Cambridge University Press.

Pinker, S. (2002). *The blank slate: The modern denial of human nature*. New York: Viking.

Rachels, J. (1990). *Created from animals: The moral implications of Darwinism*. Oxford: Oxford University Press.

Ruse, M. (1986). *Taking Darwin seriously: A naturalistic approach to philosophy*. Oxford: Blackwell.

Singer, P. (1990). *Animal liberation* (2nd edn). London: Jonathan Cape.

References

Adams, M. M. and Adams, R. M. (eds.) (1990). *The problem of evil.* Oxford: Oxford University Press.

Alexander, R. D. (1987). *The biology of moral systems.* Hawthorne, NY: Aldine de Gruyter.

Anon. (2007). Evolution and the brain. *Nature*, 447, 753.

Aquinas, T. (1948). *Summa contra gentiles* (Fathers of the English Dominican Province, trans.). New York: Benziger (original work published c. 1264).

 (1928). *Summa theologica* (Fathers of the English Dominican Province, trans.). New York: Benziger (original work published c. 1274).

Arnhart, L. (1998). *Darwinian natural right: The biological ethics of human nature.* Albany: State University of New York Press.

Atkins, P. (1993). *Creation revisited: The origin of space, time and the universe.* London: Penguin.

Atran, S. (2002). *In Gods we trust: The evolutionary landscape of religion.* Oxford: Oxford University Press.

Augustine (2003). *City of God* (H. Bettensen, trans.). London: Penguin (original work published c. 413–26).

Axelrod, R. (1984). *The evolution of cooperation.* New York: Basic Books.

Ayala, F. (1977). Nothing in biology makes sense except in the light of evolution. *Journal of Heredity*, 68, 3–10.

 (1988). Can 'progress' be defined as a biological concept? In M. H. Nitecki (ed.), *Evolutionary progress?* (pp. 75–96). Chicago: University of Chicago Press.

(2007). *Darwin's gift: To science and religion*. Washington, DC: Joseph Henry Press.

Baggini, J. (2004). *What's it all about? Philosophy and the meaning of life*. Oxford: Oxford University Press.

Barbour, I. G. (1997). *Religion and science: Historical and contemporary issues*. New York: HarperCollins.

Barkow, J. H., Cosmides, L. and Tooby, J. (eds.) (1992). *The adapted mind: Evolutionary psychology and the generation of culture*. Oxford: Oxford University Press.

Baron-Cohen, S. (1995). *Mindblindness: An essay on autism and theory of mind*. Cambridge, MA: MIT Press.

Barrett, J. L. (2004). *Why would anyone believe in God?* Lanham, MD: AltaMira Press.

Barrett, J. L. and Keil, F. C. (1996). Conceptualizing a nonnatural entity: Anthropomorphism in God concepts. *Cognitive Psychology*, 31, 219–47.

Barrett, P. H., Gautrey, P. J., Herbert, S., Kohn, D. and Smith, S. (1987). *Charles Darwin's notebooks 1836–1844: Geology, transmutation of species, metaphysical enquiries*. Ithaca, NY: Cornell University Press.

Barrett, P. H. and Gruber, H. E. (eds.) (1980). *Metaphysics, materialism, and the evolution of mind: The early writings of Charles Darwin*. Chicago: University of Chicago Press.

Barrow, J. D. and Tipler, F. J. (1986). *The anthropic cosmological principle*. Oxford: Oxford University Press.

Batson, C. D., Schoenrade, P. and Ventis, W. L. (1993). *Religion and the individual: A social-psychological perspective*. Oxford: Oxford University Press.

Behe, M. J. (1996). *Darwin's black box: The biochemical challenge to evolution*. New York: Free Press.

(2007). *The edge of evolution: The search for the limits of Darwinism*. New York: Free Press.

Beit-Hallahmi, B. (2007). Atheists: A psychological profile. In M. Martin (ed.), *The Cambridge companion to atheism* (pp. 300–17). Cambridge: Cambridge University Press.

Bering, J. (2005). The evolutionary history of an illusion: Religious causal beliefs in children and adults. In B. J. Ellis and D. F. Bjorklund (eds.), *Origins of the social mind: Evolutionary psychology and child development* (pp. 411–37). New York: Guilford Press.

Bergson, H. (1911). *Creative evolution* (A. Mitchell, trans.). New York: Henry Holt (original work published 1907).

Betzig, L. (ed.) (1997). *Human nature: A critical reader.* Oxford: Oxford University Press.

Bevc, I. and Silverman, I. (2000). Early separation and sibling incest: A test of the revised Westermarck theory. *Evolution and Human Behavior,* 21, 151–61.

Blackmore, S. (1999). *The meme machine.* New York: Oxford University Press.

Boulter, S. (2007). *The rediscovery of common sense philosophy.* Basingstoke: Palgrave Macmillan.

Bowler, P. B. (2003). *Evolution: The history of an idea* (3rd edn). London: University of California Press.

Boyd, R. and Richerson, P. J. (1985). *Culture and the evolutionary process.* Chicago: University of Chicago Press.

Boyer, P. (2001). *Religion explained: The evolutionary origins of religious thought.* New York: Basic Books.

Bradie, M. (1986). Assessing evolutionary epistemology. *Biology and Philosophy,* 1, 401–59.

Brink, D. O. (2007). The autonomy of ethics. In M. Martin (ed.), *The Cambridge companion to atheism* (pp. 149–65). Cambridge: Cambridge University Press.

Brockman, J. (ed.). (2006). *Intelligent thought: Science versus the intelligent design movement.* New York: Vintage.

Brooke, J. H. (2003). Darwin and Victorian Christianity. In J. Hodge and G. Radick (eds.), *The Cambridge companion to Darwin* (pp. 192–213). Cambridge: Cambridge University Press.

Browne, J. (2006). *Darwin's 'Origin of Species': A biography.* London: Atlantic.

Bryan, W. J. (1922). *In His image.* New York: Fleming H. Revell.

(1925/2008). Closing statement of William Jennings Bryan at the trial of John Scopes. In B. Hankins (ed.), *Evangelicalism and fundamentalism: A documentary reader* (pp. 85–95). New York: New York University Press.

Buckman, R. (2002). *Can we be good without God? Biology, behavior, and the need to believe*. New York: Prometheus.

Burnstein, E., Crandall, C. and Kitayama, S. (1994). Some neo-Darwinian decision rules for altruism: Weighing cues for inclusive fitness as a function of the biological importance of the decision. *Journal of Personality and Social Psychology*, 67, 773–89.

Buss, D. M. (1994). *The evolution of desire: Strategies of human mating*. New York: Basic Books.

(1995). Evolutionary psychology: A new paradigm for psychological science. *Psychological Inquiry*, 6, 1–30.

(2007). *Evolutionary psychology: The new science of the mind* (3rd edn). Needham Heights, MA: Allyn and Bacon.

Buss, D. M. and Schmitt, D. P. (1993). Sexual strategies theory: An evolutionary perspective on human mating. *Psychological Review*, 100, 204–32.

Butterworth, B. (1999). *The mathematical brain*. London: Macmillan.

Cairns-Smith, A. G. (1985). *Seven clues to the origin of life*. Cambridge: Cambridge University Press.

Campbell, D. T. (1974). Evolutionary epistemology. In P. A. Schilpp (ed.), *The philosophy of Karl Popper* (pp. 413–63). La Salle, IL: Open Court.

Carnegie, A. (2006). *The 'gospel of wealth' essays and other writings*. New York: Penguin.

Carruth, W. H. (1909). *Each in his own tongue, and other poems*. New York: Putnam's.

Carruthers, P. (1992). *Human knowledge and human nature: A new introduction to an ancient debate*. Oxford: Oxford University Press.

Chambers, R. (1844). *Vestiges of the natural history of creation*. London: Churchill.

Clark, A. (2008). *Supersizing the mind: Embodiment, action, and cognitive extension*. Oxford: Oxford University Press.

Clark, J. W. and Hughes, T. M. (eds.) (1890). *The life and letters of the Reverend Adam Sedgwick*, vol. II. Cambridge: Cambridge University Press.

Clark, R. and Hatfield, E. (1989). Gender differences in receptivity to sexual offers. *Journal of Psychology and Human Sexuality*, 2, 39–55.

Collins, F. S. (2006). *The language of God: A scientist presents evidence for belief.* New York: Free Press.

Conway Morris, S. (2003). *Life's solution: Inevitable humans in a lonely universe.* Cambridge: Cambridge University Press.

Cottingham, J. (2002). *On the meaning of life.* London: Routledge.

Coyne, J. A. (2009). *Why evolution is true.* New York: Viking.

Craig, W. L. and Smith, Q. (1993). *Theism, atheism, and Big Bang cosmology.* Oxford: Clarendon Press.

Crawford, C. B. and Salmon, C. (eds.) (2004). *Evolutionary psychology, public policy, and private decisions.* Mahwah, NJ: Erlbaum.

Cronin, H. (1991). *The ant and the peacock: Altruism and sexual selection from Darwin to today.* Cambridge: Cambridge University Press.

Cunningham, S. (1996). *Philosophy and the Darwinian legacy.* New York: University of Rochester Press.

Dacey, A. (2008). *The secular conscience: Why belief belongs in public life.* New York: Prometheus.

Daly, M. and Wilson, M. (1988). *Homicide.* Hawthorne, NY: Aldine de Gruyter.

Daly, M., Wilson, M. and Vasdev, S. (2001). Income inequality and homicide rates in Canada and the United States. *Canadian Journal of Criminology*, 43, 219–36.

Darwin, C. (1859). *On the origin of the species by means of natural selection.* London: Murray.

(1871). *The descent of man and selection in relation to sex.* London: Murray.

(2002). *Autobiographies.* London: Penguin.

Darwin, E. (1803). *The temple of nature, or The origin of society: A poem with philosophical notes.* London: Johnson.

Darwin, F. (ed.) (1887a). *The life and letters of Charles Darwin*, vol. i. New York: Appleton.

(1887b). *The life and letters of Charles Darwin*, vol. ii. New York: Appleton.

(1887c). *The life and letters of Charles Darwin*, vol. iii. New York: Appleton.

Darwin, F. and Seward, A. C. (eds.) (1903). *More letters of Charles Darwin. A record of his work in a series of hitherto unpublished letters*, vol. i. London: Murray.

Davies, P. (1992). *The mind of God: Science and the search for ultimate meaning*. London: Penguin.

(1999). *The fifth miracle: The search for the origin and meaning of life*. New York: Simon and Schuster.

(2006). *The Goldilocks enigma: Why is the universe just right for life?* London: Penguin.

Dawkins, M. S. (1993). *Through our eyes only? The search for animal consciousness*. Oxford: Oxford University Press.

Dawkins, R. (1982). *The extended phenotype: The long reach of the gene* (rev. edn). Oxford: Oxford University Press.

(1989). *The selfish gene* (2nd edn). Oxford: Oxford University Press.

(1995). *River out of Eden: A Darwinian view of life*. London: Weidenfeld and Nicolson.

(1998a). Human chauvinism. *Evolution*, 51, 1015–20.

(1998b). *Unweaving the rainbow: Science, delusion, and the appetite for wonder*. London: Penguin.

(2006). *The God delusion*. New York: Houghton Mifflin.

(2009). *The greatest show on earth: The evidence for evolution*. New York: Bantam Press.

Dawkins, W. B. (1871). Darwin on the Descent of Man. *Edinburgh Review*, 134, 195–235.

de Waal, F. (1996). *Good natured: The origins of right and wrong in humans and other animals*. Cambridge, MA: Harvard University Press.

(ed.) (2006). *Primates and philosophers: How morality evolved*. Princeton, NJ: Princeton University Press.

Deacon, T. W. (1997). *The symbolic species: The co-evolution of language and the brain*. New York: Norton.

Dehaene, S. (1997). *The number sense: How the mind creates mathematics*. London: Penguin.

DeLay, T. and Dawson, A. (1999). Congressional Record 6/16, H4366.

Dembski, W. A. (1998). *The design inference: Eliminating chance through small probabilities*. Cambridge: Cambridge University Press.

Dembski, W. A. and Ruse, M. (eds.) (2004). *Debating design: From Darwin to DNA*. Cambridge: Cambridge University Press.

Dennett, D. C. (1991). *Consciousness explained*. New York: Little, Brown.

(1995). *Darwin's dangerous idea: Evolution and the meanings of life.* New York: Simon and Schuster.

(2003). *Freedom evolves.* New York: Viking.

(2006). *Breaking the spell: Religion as a natural phenomenon.* New York: Viking.

Descartes, R. (1986). *Meditations on first philosophy* (J. Cottingham, trans.). Cambridge: Cambridge University Press (original work published 1641).

Desmond, A. and Moore, J. (1991). *Darwin: The life of a tormented evolutionist.* New York: Norton.

Diamond, J. (1992). *The third chimpanzee: The evolution and future of the human animal.* New York: Harper Perennial.

(1997a). *Guns, germs, and steel: The fates of human societies.* New York: Norton.

(1997b). *Why is sex fun? The evolution of human sexuality.* London: Weidenfeld and Nicolson.

Dobzhansky, T. (1973). Nothing in biology makes sense except in the light of evolution. *American Biology Teacher,* 35, 125–9.

Doolittle, R. F. (1997). A delicate balance. *Boston Review,* 22, 28–9.

Doolittle, W. F. and Zhaxybayeva, O. (2007). Reducible complexity: The case for bacterial flagella. *Current Biology,* 17, 510–12.

Dowd, M. (2007). *Thank God for evolution: How the marriage of science and religion will transform your life and our world.* New York: Viking.

Drange, T. M. (1998). *Nonbelief and evil: Two arguments for the nonexistence of God.* New York: Prometheus.

Drummond, H. (1894). *The Lowell lectures on the ascent of man.* New York: J. Pott.

Dunbar, R. I. M. and Barrett, L. (eds.) (2007). *The Oxford handbook of evolutionary psychology.* Oxford: Oxford University Press.

Dupré, J. (1992). Species: Theoretical contexts. In E. Fox Keller and E. A. Lloyd (eds.), *Keywords in evolutionary biology* (pp. 312–17). Cambridge, MA: Harvard University Press.

Durant, W. (1931). *Great men of literature.* New York: Doubleday.

Ellerbe, H. (1995). *The dark side of Christian history.* New York: Morningstar.

Ellis, B. J. and Symons, D. (1990). Sex differences in sexual fantasy: An evolutionary approach. *Journal of Sex Research*, 4, 527–55.

Fairbanks, D. J. (2007). *Relics of Eden: The powerful evidence of evolution in DNA*. New York: Prometheus.

Fehr, E. and Gaechter, S. (2002). Altruistic punishment in humans. *Nature*, 415, 137–40.

Fish, W. (2009). *Perception, hallucination, and illusion*. Oxford: Oxford University Press.

Fisher, R. A. (1947). The renaissance of Darwinism. *Listener*, 37, 1001.

Forrest, B. and Gross, P. R. (2004). *Creationism's Trojan Horse: The wedge of intelligent design*. Oxford: Oxford University Press.

Fox, R. (1967). *Kinship and marriage*. Harmondsworth: Penguin.

Freud, S. (1927). *The future of an illusion*. New York: Classic House.

Geary, D. C. (1995). Reflections of evolution and culture in children's cognition: Implications for mathematical development and instruction. *American Psychologist*, 50, 24–37.

Gonzalez, G. and Richards, J. (2004). *The privileged planet: How our place in the cosmos is designed for discovery*. Washington, DC: Regnery Publishing.

Goodenough, U. (1998). *The sacred depths of nature*. Oxford: Oxford University Press.

Gould, S. J. (1980). *The panda's thumb: More reflections in natural history*. New York: Norton.

(1996). *Full house: The spread of excellence from Plato to Darwin*. New York: Harmony.

(1999). *Rocks of ages: Science and religion in the fullness of life*. New York: Ballantine.

Grant, P. R. and Grant, B. R. (2008). *How and why species multiply: The radiation of Darwin's finches*. Princeton, NJ: Princeton University Press.

Gray, A. (1888). *Darwiniana: Essays and reviews pertaining to Darwinism*. New York: Appleton.

Gray, J. (2002). *Straw dogs: Thoughts on humans and other animals*. London: Granta.

Guthrie, S. E. (1994). *Faces in the clouds: A new theory of religion*. Oxford: Oxford University Press.

Hahlweg, K. and Hooker, C. A. (eds.) (1989). *Issues in evolutionary epistemology.* Albany: State University of New York Press.

Haidt, J. (2001). The emotional dog and its rational tail: A social intuitionist approach to moral judgment. *Psychological Review,* 108, 814–34.

Haldane, J. B. S. (1929). The origin of life. *Rationalist Annual,* 148, 3–10.

Hamer, D. H. (2004). *The God gene: How faith is hardwired into our genes.* New York: Anchor.

Hamilton, W. D. (1963). The evolution of altruistic behavior. *American Naturalist,* 97, 354–6.

(1964). The genetical evolution of social behaviour: I and II. *Journal of Theoretical Biology,* 7, 1–52.

Hamming, R. W. (1980). The unreasonable effectiveness of mathematics. *American Mathematics Monthly,* 87, 81–90.

Hare, R. M. (1989) *Essays in ethical theory.* Oxford: Oxford University Press.

Harris, S. (2004). *The end of faith: Religion, terror, and the future of reason.* New York: Norton.

Haught, J. F. (2000). *God after Darwin: A theology of evolution.* Boulder, CO: Westview Press.

Hauser, M. D. (2006). *Moral minds: How nature designed our universal sense of right and wrong.* New York: HarperCollins.

Hawking, S. W. (1988). *A brief history of time: From the Big Bang to black holes.* New York: Bantam.

Hazen, R. M. (2005). *Genesis: The scientific quest for life's origins.* Washington, DC: Joseph Henry Press.

Hick, J. (2007). *Evil and the God of love* (rev. edn). Basingstoke: Palgrave Macmillan.

Hobbes, T. (1651). *Leviathan.* Oxford: Oxford University Press.

Hodge, C. (1874). *What is Darwinism?* New York: Scribner, Armstrong.

Hofstadter, R. (1944). *Social Darwinism in American thought.* Boston: Beacon Press.

Hook, S. (1961). The atheism of Paul Tillich. In S. Hook (ed.), *Religious experience and truth: A symposium* (pp. 59–64). New York: New York University Press.

Howard-Snyder, D. (ed.). (1996). *The evidential argument from evil.* Bloomington: Indiana University Press.

Hull, D. L. (1991). God of the Galapagos. *Nature,* 352, 485–6.

(2001). *Science and selection: Essays on biological evolution and the philosophy of science.* Cambridge: Cambridge University Press.

Hume, D. (1980). *Dialogues concerning natural religion* (2nd edn). Indianapolis: Hackett (original work published 1777).

(2000). *A treatise of human nature.* Oxford: Oxford University Press (original work published 1739).

Humphrey, N. (2000). *How to solve the mind–body problem.* Bowling Green, OH: Imprint Academic.

(2006). Consciousness: The Achilles heel of Darwinism? Thank God, not quite. In J. Brockman (ed.), *Intelligent thought: Science versus the intelligent design movement* (pp. 50–64). New York: Vintage.

Hunter, C. (2001). *Darwin's God: Evolution and the problem of evil.* Grand Rapids, MI: Brazos Press.

Hursthouse, R. (1999). *On virtue ethics.* Oxford: Clarendon Press.

Huxley, T. H. (1894). *Evolution and ethics, and other essays.* London: Macmillan.

Inge, W. R. (1922). *Outspoken essays,* Series 2. New York: Longmans, Green.

Ingersoll, R. G. (2007). *The works of Robert G. Ingersoll,* vol. ii: *Lectures.* New York: BiblioBazaar.

Inlay, M. (2002). *Evolving immunity.* Retrieved 9 June 2009, from Talk Design.org. website: www.talkdesign.org/faqs/Evolving_Immunity. html.

Isaac, M. (2005). *The counter-creationism handbook.* Berkeley: University of California Press.

Johnson, P. E. (1993). *Darwin on trial.* Downers Grove, IL: InterVarsity Press.

Jordan, D. S. (1915). *War and the breed: The relation of war to the downfall of nations.* Boston, MA: Beacon Press.

Joyce, R. (2001). *The myth of morality.* Cambridge: Cambridge University Press.

Kaufman, S. (2008). *Reinventing the sacred: A new view of science, reason, and religion.* New York: Basic Books.

Kelemen, D. (2004). Are children intuitive theists? Reasoning about purpose and design in nature. *Psychological Science,* 15, 295–301.

Kenrick, D. T. and Luce, C. L. (eds.) (2004). *The functional mind: Readings in evolutionary psychology*. Boston: Allyn and Bacon.

Kenrick, D. T., Sadalla, E. K., Groth, G. and Trost, M. R. (1990). Evolution, traits, and the stages of human courtship: Qualifying the parental investment model. *Journal of Personality*, 58, 97–116.

Keynes, R. D. (ed.). (2001). *Charles Darwin's Beagle diary*. Cambridge: Cambridge University Press.

Kirkpatrick, L. A. (2005). *Attachment, evolution, and the psychology of religion*. New York: Guildford Press.

Kitcher, P. (2006). Four ways of 'biologizing' ethics. In E. Sober (ed.), *Conceptual issues in evolutionary biology* (pp. 575–86). Cambridge, MA: MIT Press.

(2007). *Living with Darwin: Evolution, design, and the future of faith*. Oxford: Oxford University Press.

Klemke, E. D. (ed.). (2000). *The meaning of life* (2nd edn). Oxford: Oxford University Press.

Krebs, D. L. (1998). The evolution of moral behaviors. In C. B. Crawford and D. L. Krebs (eds.), *Handbook of evolutionary psychology: Ideas, issues, and applications* (pp. 337–68). Mahwah, NJ: Erlbaum.

Krupp, D. B., DeBruine, L. M. and Barclay, P. (2008). A cue to kinship promotes cooperation for the public good. *Evolution and Human Behavior*, 29, 49–55.

Küng, H. (1980). *Does God exist? An answer for today*. London: Collins.

Latané, B. and Darley, J. M. (1970). *The unresponsive bystander: why doesn't he help?* New York: Appleton/Century-Croft.

Leslie, J. (1989). *Universes*. New York: Routledge.

Lewis, C. S. (1940). *The problem of pain*. New York: HarperCollins.

(1952). *Mere Christianity*. New York: HarperCollins.

Lieberman, D., Tooby, J. and Cosmides, L. (2003). Does morality have a biological basis? An empirical test of the factors governing moral sentiments relating to incest. *Proceedings of the Royal Society of London, Series B: Biological Sciences*, 270, 819–26.

(2007). The architecture of human kin detection. *Nature*, 445, 727–31.

Lorenz, K. (1982). Kant's doctrine of the a priori in the light of contemporary biology. In H. C. Plotkin (ed.), *Learning, development, and culture: Essays in evolutionary epistemology* (pp. 121–43). Chichester, UK: Wiley (original work published 1941).

Lovejoy, A. O. (1936). *The Great Chain of Being: A study of the history of an idea*. Cambridge, MA: Harvard University Press.

Lyell, C. (1835). *Principles of geology*, vol. II (3rd edn). London: Murray.

Lyell, K. M. (ed.). (1881). *Life, letters, and journals of Sir Charles Lyell*, vol. I. London: Murray.

Lynn, R. (1996). *Dysgenics: Genetic deterioration in modern populations*. Westport, CT: Praeger.

(2001). *Eugenics: A reassessment*. Westport, CT: Praeger.

Mackie, J. L. (1977). *Ethics: Inventing right and wrong*. New York: Penguin.

Maisel, E. (2009). *The atheist's way: Living well without God*. Novato, CA: New World Library.

Makalowski, W. (2007). What is junk DNA, and what is it worth? Retrieved 29 September, 2009, from Scientific American website: www.scientificamerican.com/article.cfm?id=what-is-junk-dna-and-what.

Marcus, G. (2008). *Kluge: The haphazard construction of the human mind*. New York: Houghton Mifflin.

Maritain, J. (1997). *Untrammelled approaches* (B. E. Doering, trans.). Notre Dame: University of Notre Dame Press (original work published 1973).

Maynard-Smith, J. (1964). Group selection and kin selection. *Nature, 201*, 1145–7.

Maynard Smith, J. and Szathmáry, E. (1999). *The origins of life: From the birth of life to the origins of language*. Oxford: Oxford University Press.

Mayr, E. (2001). *What evolution is*. New York: Basic Books.

McGowan, D. (ed.). (2007). *Parenting beyond belief: On raising ethical, caring kids without religion*. New York: AMACOM.

McGrath, A. (2005). *Dawkins' God: Genes, memes, and the meaning of life*. Oxford: Blackwell.

McKee, J. K., Poirier, F. E., and McGraw, W. S. (2005). *Understanding human evolution* (5th edn). Upper Saddle River, NJ: Pearson.

Mill, J. S. (1874). *Nature, the utility of religion, and theism*. London: Longmans, Green, Reader, and Dyer.

Miller, G. F. (2000). *The mating mind: How sexual choice shaped the evolution of human nature*. London: Vintage.

Miller, K. R. (1999). *Finding Darwin's God: A scientist's search for the common ground between God and evolution*. New York: Harper Perennial.

Miller, S. L. (1953). Production of amino acids under possible primitive earth conditions. *Science*, 117, 528–9.

Millgate, M. (ed.). (2001). *Thomas Hardy's public voice: The essays, speeches, and miscellaneous prose*. Oxford: Oxford University Press.

Mivart, St G. J. (1871). *On the genesis of the species*. New York: Appleton.

Monk, R. (1996). *Bertrand Russell: The spirit of solitude 1872–1921*. New York: Free Press.

Moore, A. (1891). The Christian doctrine of God. In C. Gore (ed.), *Lux Mundi* (12th edn) (pp. 41–81). London: Murray.

Moore, G. E. (1903). *Principia ethica*. Cambridge: Cambridge University Press.

Morris, H. (1972). *The remarkable birth of planet Earth*. San Diego, CA: Creation Life.

Munz, P. (1993). *Philosophical Darwinism: On the origin of knowledge by means of natural selection*. London: Routledge.

Nabokov, V. (1967). *Speak, memory: An autobiography revisited*. London: Penguin.

Nelkin, D. (2000). Less selfish than sacred? Genes and the religious impulse in evolutionary psychology. In H. Rose and S. Rose (eds.), *Alas poor Darwin: Arguments against evolutionary psychology* (pp. 17–32). London: Jonathan Cape.

Nielsen, K. (1973) *Ethics without God*. Buffalo: Prometheus.

Nietzsche, F. (1974). *The gay science: with a prelude of rhymes and an appendix of songs* (W. Kaufman, trans.). New York: Vintage (original work published 1887).

  (2005). *Thus spoke Zarathustra: A book for everyone and nobody* (Graham Parkes, trans.). Oxford: Oxford University Press (original work published 1892).

  (1982). *Daybreak: Thoughts on the prejudices of morality* (R. J. Hollingdale, trans.). Cambridge: Cambridge University Press (original work published 1881).

  (1993). *Sayings of Nietzsche*. London: Duckworth.

(2003). *Beyond good and evil* (R. J. Hollingdale, trans.). London: Penguin (original work published 1886).

Nitecki, M. H. (ed.) (1988). *Evolutionary progress?* Chicago: University of Chicago Press.

Noë, A. (2009). *Out of our heads: Why you are not your brain, and other lessons from the biology of consciousness.* New York: Hill and Wang.

Nozick, R. (1981). *Philosophical explanations.* Oxford: Clarendon Press.

(1983). About mammals and people. *New York Times Book Review*, 11.

(2001). *Invariances: The structure of the objective world.* Cambridge, MA: Belknap Press.

O'Hear, A. (1997). *Beyond evolution: Human nature and the limits of evolutionary explanation.* Oxford: Clarendon Press.

Oparin, A. I. (1924). *The origin of life.* Moscow: Moscow Worker.

Paley, W. (1802). *Natural theology: Or, evidences of the existence and attributes of the deity, collected from the appearances of nature.* Oxford: Oxford University Press.

Parfit, D. (1984). *Reasons and persons.* Oxford: Clarendon Press.

Pascal, B. (1910). *Thoughts, letters, and minor works.* The Harvard Classics, vol. xlviii. New York: Collier.

Patterson, C. (2002). *Eternal Treblinka: Our treatment of animals and the Holocaust.* New York: Lantern.

Paul, D. B. (2003). Darwin, social Darwinism and eugenics. In J. Hodge and G. Radick (eds.), *The Cambridge companion to Darwin* (pp. 214–39). Cambridge: Cambridge University Press.

Paul, G. S. (2005). National correlations of quantifiable societal health with popular religiosity and secularism in the prosperous democracies. A first look. *Journal of Religion and Society*, 7, 1–17.

Peacocke, A. (2004). *Evolution: The disguised friend of faith?* Philadelphia, PA: Templeton Foundation Press.

Pennock, R. T. (1999). *Tower of Babel: The evidence against the new creationism.* Cambridge, MA: MIT Press.

Pigliucci, M. and Kaplan, J. (2006). *Making sense of evolution: The conceptual foundations of evolutionary biology.* Chicago: University of Chicago Press.

Pinker, S. (1994). *The language instinct: The new science of language and mind*. London: Penguin.

(1997). *How the mind works*. London: Penguin.

Pinker, S. and Bloom, P. (1992). Natural language and natural selection. In J. H. Barkow, L. Cosmides and J. Tooby (eds.), *The adapted mind: Evolutionary psychology and the generation of culture* (pp. 451–93). New York: Oxford University Press.

Plantinga, A. (1974). *God, freedom, and evil*. New York: Harper and Row.

Plato (1992). *Republic* (G. M. A. Grube, trans.). Indianapolis, IN: Hackett.

Plotkin, H. C. (1993). *Darwin machines and the nature of knowledge*. Cambridge, MA: Harvard University Press.

Polkinghorne, J. (1994). *Science and Christian belief: Theological reflections of a bottom-up thinker*. London: SPCK.

Popper, K. (1987). Natural selection and the emergence of mind. In G. Radnitzky and W. W. Bartley III (eds.), *Evolutionary epistemology, rationality, and the sociology of knowledge* (pp. 139–55). La Salle, IL: Open Court.

Pratt, A. R. (ed.) (1994). *The dark side: Thoughts on the futility of life from the ancient Greeks to the present*. New York: Citadel Press.

Prothero, D. R. (2007). *Evolution: What the fossils say and why it matters*. New York: Columbia University Press.

Rachels, J. (1990). *Created from animals: The moral implications of Darwinism*. Oxford: Oxford University Press.

Radcliffe Richards, J. (2000). *Human nature after Darwin*. London: Routledge.

Radnitzky, G. and Bartley III, W. W. (eds.) (1987). *Evolutionary epistemology, rationality, and the sociology of knowledge*. La Salle, IL: Open Court.

Railton, P. (1986). Moral realism. *The Philosophical Review*, 95, 163–207.

Ramsey, F. P. (2001). *The foundations of mathematics, and other logical essays*. London: Routledge.

Rawls, J. (1971). *A theory of justice*. Cambridge, MA: Harvard University Press.

Rees, M. (1997). *Before the beginning: Our universe and others*. London: Free Press.

(1999). *Just six numbers: The deep forces that shape the universe*. London: Phoenix.

Regan, T. (1983). *The case for animal rights*. California: University of California Press.

Richards, R. (1986). A defense of evolutionary ethics. *Biology and Philosophy*, 1, 265–93.

Richards, R. J. (1999). Darwin's romantic biology: The foundation. In J. Maienschein and M. Ruse (eds.), *Biology and the foundation of ethics* (pp. 113–53). Cambridge: Cambridge University Press.

Richerson, P. J. and Boyd, R. (2004). *Not by genes alone: How culture transformed human evolution*. Chicago: University of Chicago Press.

Ridley, M. (1994). *The red queen: Sex and the evolution of human nature*. London: Penguin.

(1996). *The origins of virtue: Human instincts and the evolution of cooperation*. New York: Penguin.

Rose, S. (1978). Pre-Copernican sociobiology? *New Scientist*, 80, 45–46.

Rosenberg, A. (2000). *Darwinism in philosophy, social science, and policy*. Cambridge: Cambridge University Press.

(2003). Darwinism in moral philosophy and social theory. In J. Hodge and G. Radick (eds.), *The Cambridge companion to Darwin* (pp. 310–32). Cambridge: Cambridge University Press.

Ross, H. (2001). *The creator and the cosmos: How the latest scientific discoveries reveal God* (3rd edn). Colorado Springs, CO: NavPress.

Roughgarden, J. (2006). *Evolution and Christian faith: Reflections of an evolutionary biologist*. Washington, DC: Island Press.

Royce, J. (1955). *The spirit of modern philosophy* (2nd edn). Boston: Houghton Miffin (original work published 1892).

Ruse, M. (1986). *Taking Darwin seriously: A naturalistic approach to philosophy*. Oxford: Blackwell.

(1988). *Philosophy of biology today*. Albany: State University of New York Press.

(1995). *Evolutionary naturalism: Selected essays*. London: Routledge.

(1997). *Monad to man: The concept of progress in evolutionary biology*. Cambridge, MA: Harvard University Press.

(2001). *Can a Darwinian be a Christian? The relationship between science and religion.* Cambridge: Cambridge University Press.

Ruse, M. and Wilson, E. O. (2006). Moral philosophy as applied science. In E. Sober (ed.), *Conceptual issues in evolutionary biology* (pp. 555–73). Cambridge, MA: MIT Press.

Russell, B. (1935). *Religion and science.* London: Oxford University Press.

(1950). *Unpopular essays.* London: Routledge.

(1953). *Satan in the suburbs, and other stories.* New York: Simon and Schuster.

(1957). *Why I am not a Christian, and other essays on religion and related subjects.* New York: Simon and Schuster.

Sagan, C. (1985). *Contact.* New York: Simon and Schuster.

Schaller, G. B. (1972). *The Serengeti lion: A study of predator–prey relations.* Chicago: University of Chicago Press.

Schellenberg, J. L. (1993). *Divine hiddenness and human reason.* Ithaca, NY: Cornell University Press.

Schmitt, D. P. (2005). Sociosexuality from Argentina to Zimbabwe: A 48-nation study of sex, culture, and strategies of human mating. *Behavioral and Brain Sciences*, 28, 247–75.

Schrödinger, E. (1944). *What is life?* Cambridge: Cambridge University Press.

Scruton, R. (2000). *Animal rights and wrongs* (3rd edn). London: Demos.

Segerstrale, U. (2000). *Defenders of the truth: The sociobiology debate.* Oxford: Oxford University Press.

Seyfarth, R. M. and Cheney, D. L. (1984). Grooming alliances and reciprocal alliances in vervet monkeys. *Nature*, 308, 541–3.

Shepher, J. (1971). Mate selection among second generation kibbutz adolescents and adults: Incest avoidance and negative imprinting. *Archives of Sexual Behavior*, 1, 293–307.

(1983). *Incest: A biosocial view.* New York: Academic Press.

Sherman, P. W. (1977). Nepotism and the evolution of alarm calls. *Science*, 197, 1246–53.

Shermer, M. (2006). *Why Darwin matters: The case against intelligent design.* New York: Times.

Shubin, N. (2008). *Your inner fish: A journey into the 3.5 billion-year history of the human body.* New York: Pantheon.

Simpson, G. G. (1967). *The meaning of evolution: A study of the history of life and of its significance for man* (rev. edn). New Haven, CT: Yale University Press.

Singer, I. B. (1968). *The séance and other stories*. New York: Farrar, Straus and Giroux.

Singer, P. (1981). *The expanding circle: Ethics and sociobiology*. New York: Farrar, Strauss and Giroux.

(1993). *Practical ethics* (2nd edn). Cambridge: Cambridge University Press.

(2000). *A Darwinian left*. New Haven, CT: Yale University Press.

(2002a). *Unsanctifying human life: Essays on ethics*. Oxford: Blackwell.

(2002b). *Animal liberation* (3rd edn). New York: HarperCollins.

Smith, G. H. (1980). *Atheism: The case against God*. New York: Prometheus.

Smolin, L. (1992). Did the universe evolve? *Classical and Quantum Gravity*, 9, 173–91.

(1997). *The life of the cosmos*. Oxford: Oxford University Press.

Sober, E. (2000). *Philosophy of biology* (2nd edn). Boulder, CO: Westview.

(ed.) (2006). *Conceptual issues in evolutionary biology* (3rd edn). Cambridge, MA: MIT Press.

Sober, E. and Wilson, D. S. (1998). *Unto others: The evolution and psychology of unselfish behavior*. Cambridge, MA: Harvard University Press.

Sociobiology Study Group for the People (1975). Against sociobiology. *New York Review of Books*, 22, 43–44.

Spencer, H. (1868). *Essays: Scientific, political, and speculative*, vol. ii. London: Williams and Norgate.

(1874). *The study of sociology*. London: Williams and Norgate.

Spiegel, M. (1996). *The dreaded comparison: Human and animal slavery*. New York: Mirror.

Stamos, D. M. (2004). *The species problem: Biological species, ontology, and the metaphysics of biology*. Lanham, MD: Lexington.

Stenger, V. J. (2003). *Has science found God? The latest results in the search for purpose in the universe*. New York: Prometheus.

(2007). *God, The failed hypothesis: How science shows that God does not exist*. New York: Prometheus.

Sterelny, K. (2003). *Thought in a hostile world: The evolution of human cognition.* Oxford: Blackwell.

Sterelny, K. and Griffiths, P. E. (1999). *Sex and death: An introduction to philosophy of biology.* Chicago: University of Chicago Press.

Stewart-Williams, S. (2002). Life after death. *Philosophy Now*, 39, 22–5.

(2003). Life from non-life: Must we accept a supernatural explanation? *The Skeptic*, 16, 12–16.

(2004a). Life after Darwin: Human beings and their place in the universe. *Anthropology and Philosophy*, 5, 37–47.

(2004b). Darwin meets Socrates: The implications of evolutionary theory for ethics. *Philosophy Now*, 45, 26–9.

(2004c). Can an evolutionist believe in God? *Philosophy Now*, 47, 19–21.

(2005). Innate ideas as a naturalistic source of metaphysical knowledge. *Biology and Philosophy*, 20, 791–814.

(2007). Altruism among kin vs. nonkin: Effects of cost of help and reciprocal exchange. *Evolution and Human Behavior*, 28, 193–8.

(2008). Human beings as evolved nepotists: Exceptions to the rule and effects of cost of help. *Human Nature*, 19, 414–425.

Strawson, G. (1986). *Freedom and belief.* Oxford: Clarendon Press.

Sumner, W. G. (1883). *What social classes owe to each other.* New York: Harper.

Susskind, L. (2006). *The cosmic landscape: String theory and the illusion of intelligent design.* New York: Little, Brown.

Swinburne, R. (1993). *The coherence of theism* (rev. edn). Oxford: Clarendon Press.

(1997). *The evolution of the soul* (rev. edn). Oxford: Clarendon Press.

(1998). *Providence and the problem of evil.* Oxford: Clarendon Press.

(2002). *The justification of theism.* Retrieved 20 October 2009, from Leadership University: Truth Journal website: www.leaderu.com/truth/3truth09.html.

(2004). *The existence of God* (2nd edn). Oxford: Clarendon Press.

Symons, D. (1979). *The evolution of human sexuality.* Oxford: Oxford University Press.

Teilhard de Chardin, P. (1959). *The phenomenon of man* (B. Wall, trans.). New York: HarperCollins (original work published 1955).

Tennessen, H. (1973). Knowledge versus survival. *Inquiry*, 16, 407–14.

Theobald, D. (2007). *29+ evidences for macroevolution: The scientific case for common descent, Version 2.87*. Retrieved 20 October 2009, from TalkOrigins Archive: Exploring the Creation/Evolution Controversy website: www.talkorigins.org/faqs/comdesc/.

Thornhill, R. (1998). Darwinian aesthetics. In C. B. Crawford and D. L. Krebs (eds.), *Handbook of evolutionary psychology: Ideas, issues, and applications* (pp. 543–72). Mahwah, NJ: Erlbaum.

Thornhill, R. and Palmer, C. T. (2000). *A natural history of rape: Biological bases of sexual coercion*. Cambridge, MA: MIT Press.

Tipler, F. J. (1994). *The physics of immortality: Modern cosmology, God, and the resurrection of the dead*. New York: Doubleday.

Tomasello, M. and Call, J. (1997). *Primate cognition*. Oxford: Oxford University Press.

Tooby, J. and Cosmides, L. (1992). The psychological foundations of culture. In J. H. Barkow, L. Cosmides and J. Tooby (eds.), *The adapted mind: Evolutionary psychology and the generation of culture* (pp. 19–136). New York: Oxford University Press.

Trevor-Roper, H. (ed.) (1963). *Hitler's table-talk*. London: Weidenfeld and Nicolson.

Trivers, R. L. (1971). The evolution of reciprocal altruism. *Quarterly Review of Biology*, 46, 35–57.

  (1985). *Social evolution*. Menlo Park, CA: Benjamin/Cummings.

  (2002). *Natural selection and social theory: Selected papers of Robert Trivers*. Oxford: Oxford University Press.

Tryon, E. P. (1973). Is the universe a vacuum fluctuation? *Nature*, 246, 396–7.

Twain, M. (1924). *Mark Twain's autobiography*, vol. ii. New York: Harper.

  (1992). *Collected tales, sketches, speeches, and essays, vol. ii: 1891–1910*. New York: Library of America.

Van den Berghe, P. L. (1990). From the Popocatepetl to the Limpopo. In B. M. Berger (ed.), *Authors of their own lives: Intellectual autobiographies by twenty American sociologists*. Berkeley: University of California Press.

Vilenkin, A. (1982). The creation of universes from nothing. *Physical Letters*, 117B, 25–8.

—— (2006). *Many worlds in one: The search for other universes.* New York: Hill and Wang.

Wallace, A. R. (1870). *Contributions to the theory of natural selection: A series of essays.* London: Macmillan.

Ward, K. (1996). *God, chance, and necessity.* Oxford: Oneworld.

—— (2006). *Is religion dangerous?* Oxford: Lion Hudson.

Warren, R. (2002). *The purpose-driven life: What on earth am I here for?* Grand Rapids, MI: Zondervan.

Weikart, R. (2004). *From Darwin to Hitler: Evolutionary ethics, eugenics, and racism in Germany.* New York: Palgrave Macmillan.

Weiner, J. (1994). *The beak of the finch: A story of evolution in our time.* New York: Vintage.

Weisberger, A. M. (2007). The argument from evil. In M. Martin (ed.), *The Cambridge companion to atheism* (pp. 166–81). Cambridge: Cambridge University Press.

Westermarck, E. A. (1921). *The history of human marriage* (5th edn). London: Macmillan.

White, A. D. (1896). *A history of the warfare of science with theology in Christendom.* New York: Appleton.

Whiten, A. (1997). The Machiavellian mindreader. In A. Whiten and R. W. Byrne (eds.), *Machiavellian intelligence II: Extensions and evaluations* (pp. 144–73). Cambridge: Cambridge University Press.

Wigner, E. P. (1960). The unreasonable effectiveness of mathematics in the natural sciences. *Communications on Pure and Applied Mathematics*, 13, 1–14.

Wilkinson, G. S. (1984). Reciprocal food sharing in the vampire bat. *Nature*, 308, 181–4.

Williams, B. (1972). *Morality: An introduction to ethics.* New York: Harper and Row.

Williams, G. C. (1966). *Adaptation and natural selection: A critique of some current evolutionary thought.* Princeton, NJ: Princeton University Press.

—— (1993). Mother Nature is a wicked old witch. In M. H. Nitecki and D. V. Nitecki (eds.), *Evolutionary ethics* (pp. 217–31). Albany: State University of New York Press.

Wilson, D. S. (2002). *Darwin's cathedral: Evolution, religion, and the nature of society.* Chicago: University of Chicago Press.

Wilson, D. S., Dietrich, E., and Clark, A. B. (2003). On the inappropriate use of the naturalistic fallacy in evolutionary psychology. *Biology and Philosophy*, 18, 669–82.

Wilson, D. S. and Sober, E. (1994). Reintroducing group selection to the human behavioral sciences. *Behavioral and Brain Sciences*, 17, 585–654.

Wilson, E. O. (1975). *Sociobiology: The new synthesis*. Cambridge, MA: Belknap Press.

(1978). *On human nature*. Cambridge, MA: Harvard University Press.

(1984). *Biophilia*. Cambridge, MA: Harvard University Press.

(1996). *In search of nature*. London: Penguin.

(1998). *Consilience: The unity of knowledge*. New York: Knopf.

(1999). *The diversity of life* (rev. edn). New York: Norton.

(2002). *The future of life*. New York: Knopf.

Wittgenstein, L. (1921). *Tractatus logico-philosophicus*. London: Routledge.

Wolf, A. P. (1995). *Sexual attraction and childhood association: A Chinese brief for Edward Westermarck*. Stanford, CA: Stanford University Press.

Wright, R. (1994). *The moral animal: The new science of evolutionary psychology*. New York: Pantheon.

Zahavi, A. (1975). Mate selection: A selection for a handicap. *Journal of Theoretical Biology*, 53, 205–14.

Zuckerman, P. (2007). Atheism: Contemporary numbers and patterns. In M. Martin (ed.), *The Cambridge companion to atheism* (pp. 47–65). Cambridge: Cambridge University Press.

(2008). *Society without God: What the least religious nations can tell us about contentment*. New York: New York University Press.

Index